Regional Growth and Regional Policy Within the Framework of European Integration

Karin Peschel

Karin Peschel (Ed.)

Regional Growth and Regional Policy Within the Framework of European Integration

Proceedings of a Conference on the Occasion of 25 Years Institute for Regional Research at the University of Kiel 1995

With 17 Figures

Physica-Verlag
A Springer-Verlag Company

Series Editors
Werner A. Müller
Peter Schuster

Editor
Prof. Dr. Karin Peschel
Institute for Regional Research
University of Kiel
Olshausenstraße 40
D-24098 Kiel, Germany

Editorial Supervisor
Markus Hirschfeld

ISBN 978-3-7908-0957-2 ISBN 978-3-642-48428-5 (eBook)
DOI 10.1007/978-3-642-48428-5

Die Deutsche Bibliothek – CIP-Einheitsaufnahme
Regional growth and regional policy within the framework of European integraton: proceedings of a Conference on the Occasion of 25 Years Institute for Regional Research at the University of Kiel 1995/Karin Peschel (ed.) – Heidelberg: Physica-Verl., 1997
(Contributions to economics)

NE: Peschel, Karin [Hrsg.]; Conference on the Occasion of 25 Years Institute for Regional Research at the University of Kiel <1995, Kiel>; Institut für Regionalforschung <Kiel>

SPIN 10544030 88/2202-5 4 3 2 1 0 – Printed on acid-free paper

Preface

The twenty-fifth anniversary of the Institute for Regional Research (*Institut für Regionalforschung*) was celebrated by an international conference on

"Regional Growth and Regional Policy within the Framework of European Integration".

We are happy to be able to present the outcome of that conference in this volume. Its authors reflect the Institute's international character. Their contributions characterize main fields of interest of the Institute's research staff.

The Institute for Regional Research is an establishment of the Economic and Social Science Faculty (*Wirtschafts- und Sozialwissenschaftliche Fakultät*) at Christian-Albrechts-University in Kiel, one of the faculty's four institutes of economics. It began as an initiative of Reimut Jochimsen. He wanted to upgrade regional research and planning at the university because at the faculty level this field was only a minor province. As further justification for the foundation of an institute, Jochimsen referred to the growing international and national significance of regional research and especially of regional structural policy. The need for an autonomous institute in the north of Germany was emphasized above all, particularly since regional science in the Federal Republic of Germany in the late sixties was largely limited to the universities in Freiburg, Karlsruhe and Münster. It was emphasized as well, that an institute for regional research in Schleswig-Holstein would be to the state's advantage as it could help the state take policy decisions on regional political or economic measures on the basis of continuous study of the regions. When the Institute was finally founded in 1970, Reimut Jochimsen was on temporary leave from Kiel due to his activities at the Federal Chancellery in Bonn. That is why within the subject theoretical economics Karin Peschel had delivered lectures in regional economics (*Raumwirtschaftstheorie*) beginning with the 1972/73 winter term. It was not until the 1977/78 winter term that students could choose regional science as an optional course within their curriculum. The supervision of students was largely based on the voluntary work of externally funded scientists. Thanks to their efforts, well over 100 diploma theses (masters) have been written at the Institute up to now.

In 1974, Karin Peschel became head of the Institute together with Manfred Willms, under very adverse conditions: through many years the Institute had to work without a permanent assistant's post, that is, it existed only with externally financed scientists. Only due to the special research area (*Sonderforschungsbereich)* "Research for the Scandinavian and Baltic Sea Region" with several scientists engaged in longer-term research projects, could the Institute prevent rapid turnover of the scientific team, gain the necessary international reputation and ensure its existence. As most important staff members are to be named Johannes Bröcker, whom the Institute owes its high standard in theory and econometrics, Walter Reimers and Jens Haass. After the *Sonderforschungsbereich* ended in 1982/83, it still took ten more years before, in 1993, a permanent post for the management of the Institute's data base and for ongoing regional analyses for Schleswig-Holstein was finally installed. This post - occupied since that time by Hayo Herrmann - allows ongoing work in the field of empirical research on Schleswig-Holstein and Northern Germany.

Despite the wide spectrum of empirical regional research projects that have characterized the Institute's work over almost 25 years, three main subject areas can be identified:

- changes in international and interregional division of labor and regional growth differences resulting from long-term integration processes in West Europe, paying special attention to Northern Europe (Scandinavia), and recently the entire Baltic Sea area.
- determinants of regional differences in growth and structural change in Germany.
- structural change and growth differences in Schleswig-Holstein and in its regions - analysis of regional labour markets and interlinks.

These subjects are well represented in recent research projects as well as in continued empirical work at the Institute and in the dissertations of its members. The first subject mentioned, "integration and interregional division of labor", is a main interest of Karin Peschel. This field has considerably determined the Institute's integration into Northern European and Scandinavian research. The empirical work on Schleswig-Holstein, of which Hayo Herrmann is in charge, has grown into a further core activity in recent years - more or less in counterpoint to the emphasis on European research. The Institute has thus become an established part of Schleswig-Holstein's research in regional empirical analyses.

The fact that Karin Peschel is a founder member of the German Section of the Regional Science Association (RSA) and that she was in the organizing committee of the international RSA for many years led to a multitude of contacts. In the course of time there have arisen excellent connections with scientists in the Scandinavian and Baltic Sea regions: first from 1974 to 1982 due to the projects in the *Sonderforschungsbereich* "Research for the Scandinavian and Baltic Sea Region", and later due to good connections within the "Nordic Section" of the Regional Science Association. Recently connections are due to heightened co-operation with scientists from Poland, the Baltic States and Russia.

Among the scientists and institutions, with which the Institute had co-operated are, for example:

- Åke E. Anderson, Institute for Futures Studies, Stockholm,
- Folke Snickars and Börje Johannson, Royal Institute of Technology, Stockholm,
- Nils Groes, AKF, Amternes og Kommunernes Forskningsinstitut, Copenhagen,
- Arild Sæther, Department of Economics, Agder State College, Kristiansand,
- Institut for Grænseregionsforskning, Aabenraa (the *Sonderforschungsbereich* mentioned above contributed much to the founding of this institute),
- Henryk Cwiklinski, Department of Economic Policy, Faculty of Economics, University of Gdansk,
- Murat Albegov, International University, Moskow.

These contacts have led to some present projects, other projects are still being planned.

In the German-speaking regions the Institute is closely connected with Rolf Funck and Werner Rothengatter, *Institut für Wirtschafts- und Sozialwissenschaften*, University of Karlsruhe, with Franz-Josef Bade, *Institut für Raumwirtschaftspolitik*, University of Dortmund, and of course with Johannes Bröcker, head of the *Institut für Wirtschaft und Verkehr*, University of Dresden, since 1993, who determined the Institute's work for a long time. Of course, there are regular contacts with all important German institutions working in the field of empirical regional research, in particular the *Institut für Arbeitsmarkt- und Berufsforschung* in Nürnberg, the *Akademie für Raumforschung und Landesplanung* in Hannover and the *Bundesforschungsanstalt für Landeskunde und Raumordnung* in Bonn.

In 1971 the Institute's first publications - written in English - came out under the designation "*Discussion Paper*". This serial has been continued since 1980 as "*Diskussionsbeiträge aus dem Institut für Regionalforschung*", appearing originally in English and later in German. Other publications are the "*Beiträge aus dem Institut für Regionalforschung*" (since 1982) and the serial "*Schriften des Instituts für Regionalforschung der Universität Kiel*" (published by V. Florentz, Munich). All publications are edited by Karin Peschel.

For publication of this conference volume we would like to thank the President of the *Landeszentralbank in der Freien und Hansestadt Hamburg, in Mecklenburg-Vorpommern und Schleswig-Holstein,* Prof. Dr. Hans-Jürgen Krupp.

Kiel, May 1996

Karin Peschel

Contents

Regional Science in Germany
- From the Beginnings to the Present -

Reimut Jochimsen
President of the Landeszentralbank in Nordrhein-Westfalen, Düsseldorf and Member of the Central Bank Council of the Deutsche Bundesbank, Frankfurt/M.
Berliner Allee 14
D - 40212 Düsseldorf

When Karin Peschel approached me to ask if I would provide a talk on the development of regional science in Germany to mark the "silver anniversary" of the Institute of Regional Research here in Kiel, I accepted quite spontaneously and gladly. And I will make no secret of the fact that I consider myself honoured to have been asked. For although I was the founder of the institute two-and-a-half decades ago, in reality I had to renounce its directorship quite soon afterwards (1976), when I moved from the Baltic to take up an appointment in the Federal Chancellery in Bonn (1970 - 1973) and later the Federal Ministry of Education and Science, subsequently moving to Düsseldorf (1978) to head the North-Rhine Westphalian Science and Economics Ministries (till 1990).

Since that time, though, regional research has never been quite out of my mind during these various offices; nor is it now that I have turned into a central banker. On the contrary, even after 20 years in the "hot seats" of political office and the quite dramatic shift from federal government to the *Land* of North-Rhine-Westphalia, the roots of my own academic studies have never lost their fascination for me, and never lost my involvement although - to my great regret - the time available has always been sparse, so I cannot claim to have a full, up-to-the-minute picture of our discipline at my finger-tips. However, I did have an interesting opportunity 15 years ago to offer some reflections on the paradigm shift within our discipline to the *Akademie für Raumforschung und Landesplanung* at Osnabrück in 1980[1]; I addressed the same body of people again in Aachen in 1991[2],

---

[1] Jochimsen, Reimut: Regionalpolitik und Raumordnung - über die Verwirklichung als politische Aufgabe. In: Akademie für Raumforschung und Landesplanung (ed.), Räumliche Planung in der Bewährung, Wissenschaftliche Plenartagung Osnabrück 1980, Hanover 1982, pp. 5-19.

[2] Jochimsen, Reimut: Finanz- und wirtschaftspolitische Rahmenbedingungen der europäischen Integration. In: Akademie für Raumforschung und Landesplanung (ed.), Perspektiven der Raumentwicklung in Europa, Wissenschaftliche Plenartagung Aachen 1991, Hanover 1992, pp. 40-61.

when my brief was to analyse the tremendous practical advances made in integrating regional economies into a larger Europe, the prospects and challenges against a background of dissolving boundaries between national economies around the world, the break-down of the Berlin Wall and the Eastern Block, the reunification of Germany and the political programme of integration in the European Community context. Then in 1992 in Weimar[3], I endeavoured to fit the infrastructure tasks facing the new *Länder* into this overall picture. Later, in 1993, in three papers for a Hanover Conference[4], for the Federal Ministry for Regional Planning, Building and Urban Development[5] and for the Academic Memorial Adress for my academic teacher, J. Heinz Müller, at the University of Freiburg[6], and 1994 in a paper for the *Institut für Landes- und Stadtentwicklungsforschung*[7] to honour Viktor v. Malchus, its retiring director, I sought to round out the picture.

Turning to the occasion of the silver jubilee of the Kiel Institute for Regional Research, first of all, can I say what a pleasure it is to see that the IfR has established itself today as a both nationally and internationally reputed member of the scientific community in the complex field of regional research. My heartfelt congratulations to you for this splendid achievement. For things have certainly come a long way during these years in regional science in Germany, and here in Kiel in particular.

---

[3] Jochimsen, Reimut: Thesen zu den räumlichen Erfordernissen einer Infrastrukturpolitik für das Jahr 2000. In: Akademie für Raumforschung und Landesplanung (ed.), Infrastrukturelle Voraussetzungen des Strukturwandels, Wissenschaftliche Plenartagung Weimar 1992, Hanover 1993, pp. 58-69.

[4] Jochimsen, Reimut: Finanzausgleich im Bundesstaat: Europäische Perspektiven und ihre Implikationen für den Finanzausgleich in der Bundesrepublik. In: Staatswissenschaften und Staatspraxis, No. 1, 1993, pp. 118-135.

[5] Jochimsen, Reimut: Raumentwicklung in Deutschland - Deutsche Einheit und europäischer Einigungsprozeß als neue Herausforderung für Politik, Wirtschaft und Umwelt. In: Bundesforschungsanstalt für Landeskunde und Raumordnung (ed.), Raumentwicklung. Politik für den Standort Deutschland, Bonn 1993, pp. 3-13.

[6] Jochimsen, Reimut: Raumordnung und regionale Strukturpolitik - Anmerkungen zu einem neuen Spannungsfeld im vereinten Deutschland, Gedenkrede auf J. Heinz Müller, Freiburger Universitätsblätter, Freiburg, Germany 1994. Reprinted in: Jahrbuch der Görres Gesellschaft, 1994.

[7] Jochimsen, Reimut: Herausforderungen für die Raumentwicklungspolitik im vereinten Europa. In: ILS-Schriften, 94, pp. 12-17.

1 The Early Roots

In the late 1960s, as the first plans to establish this institute came to maturity, regional research in Germany was still in the midst of its growing-up phase. That meant it needed to acquire many of the facets which are part-and-parcel of the scientific organizational style of an established research programme in a rivalry with other fields.

Even in the post-war period, the roots of regional science were still spread widely among quite different disciplines. On the one hand, some vital foundations were provided by classical spatial sciences, headed by geography. On the other, a number of originally "non-spatial" sciences soon discovered the regional dimension as an indispensable addition to their fields of research. Economics, in particular, had devoted attention at an early stage to the structural problems of choosing locations, and to the links between wider-ranging trade, international transport costs and national or regional specialization.

Many of these early contributions came from the field of economics (i.e., *Nationalökonomie*) in Germany, which can boast a long, respected tradition in the study of locations in particular: We may recall Johann Heinrich von Thünen's *"Der isolirte Staat"*,[8] which provided the very first location-theoretical approach to the spatial attachment of economic processes way back in 1826, and also Alfred Weber, who founded the specific theoretical analysis of industrial locations in 1909,[9] after Wilhelm Roscher and Albert Schäffle had been the first to scrutinize the regional dimension of the industrialization process empirically.

Even well into the inter-war period, it was fair to say that German researchers exercised a virtual hegemony in this field. It was Tord Palander, the great Swedish regional economist, who observed in 1935 that all of the most important enquiries on the theory of industrial location had originated in Germany.[10]

Other works which one must not on any account omit from this sketch of the early days of regional research in Germany are, for one, Walter Christaller's (1933)[11] contribution on the significance of central functions in the development of settlement patterns and, for another, August Lösch's famous market networks and the explanatory patterns of spatial structure deduced from them (1939).[12]

---

[8] Thünen, Johann Heinrich von: Der isolirte Staat in Beziehung auf Landwirtschaft und Nationalökonomie, (I) Hamburg 1826.

[9] Weber, Alfred: Über den Standort der Industrien, Part 1, Tübingen 1909.

[10] Palander, Tord: Beiträge zur Standortstheorie, Upsala 1935, p. 137.

[11] Christaller, Walter: Die zentralen Orte in Süddeutschland, Jena 1933.

[12] Lösch, August: Die räumliche Ordnung der Wirtschaft, Jena 1939.

Right up to the present day, all of these works occupy a central place in regional economics when it comes to explaining agricultural, industrial and service locations, transportation links between them, and also the associated conurbations and settlement structures. In spite of these splendid individual achievements, there can be no overlooking the fact that the explanations available up to and into the 1950s were still only partial approaches from a theoretical viewpoint, while on the empirical side people were mainly busily engaged in collecting together rather isolated facts.

An interactive approach was still lacking well into the post-war period. Andreas Predöhl noted in 1959 that: "The only person to have endeavoured to establish a theory of spatial ordering as a whole is August Lösch."[13] Although Lösch was not able to bring his intentions to fruition, he had clearly seen the need to reach out from the partial approach of location theory to develop a spatial economic theory. In his *Die räumliche Ordnung der Wirtschaft* ("Spatial Order in the Economy"), he noted: "Once again we have a shift of emphasis in economic theory: initially the focus was on a concept of price theory isolated from time and space, then interest-rate and cyclical theories at least took account of time, and now we have embarked on the third period - we are taking the spatial aspect seriously." He wrote this in the early 1940s. So Lösch had started a general theory on integration effects but could not complete it.

2 The "Quantum Leap" to Regional Science

The "take-off" in which these scattered approaches really were merged into an autonomous discipline did not occur until the post-war period. Herbert Giersch, on the one hand, was one of the researchers who contributed in 1950 a methodologically important paper on the explanation of distributional consequences of locational structures.[14] One of the researchers in Germany to take up the various specific theories available was Edwin von Böventer, who built on them to develop pioneering approaches to an integrated explanation of spatial structure, basing the explanations on exogenous *raumdifferenzierende Faktoren.*[15] Regional economic

---

[13] Predöhl, Andreas: General discussion on Interregionale Beziehungen wirtschaftlicher Aktivitäten. In: Probleme des räumlichen Gleichgewichtes in der Wirtschaftswissenschaft, Schriften des Vereins für Socialpolitik, N.F. 14, Berlin 1959, p. 62.

[14] Giersch, Herbert: Probleme der regionalen Einkommensverteilung. In: Probleme des räumlichen Gleichgewichtes ..., op. cit.

[15] Böventer, Edwin von: Die Struktur der Landschaft. Versuch einer Synthese und Weiterentwicklung der Modelle J.H. v.Thünens, W. Christallers und A. Löschs. In: Optimales Wachstum und optimale Standortverteilung, Berlin 1962; Böventer: Theorie des räumlichen Gleichgewichts, Tübingen 1962.

research became increasingly interlinked in methodological terms with the theories of foreign trade, transport costs and transport sciences, and the theories of economic integration in general (Eli F. Heckscher, Bertil G. Ohlin, Bela Balassa).

In the USA, Walter Isard formulated his first proposals during the 1950s for linking together the regional economic models which had so far been mutually isolated; his pioneering *Channels of Synthesis* followed in 1960. Moreover, he organized a new scientific platform in the shape of the Regional Science Association, aiming to put forward new ideas towards creating a new science, and to establish the complex relations between humanity and space as a new object of research. This was the true cradle of regional science.

From the mid-1950s, the new programme of knowledge and research was clearly asserting itself: "regional science" as an integrative research framework for the complex, geographically differentiated economic system, particularly in the developed and newly developing industrial countries. As it saw its own role, this new regional science not only encompassed the general microeconomic theory of equilibrium but also certain aspects of the planning sciences, administrative studies, politics, transport sciences, and more recently also increasingly ties with ecology and evolutionary economics. Given the complex nature of the subject, this fundamentally interdisciplinary approach that was a characteristic principle of the advancement of knowledge and research in regional science was both necessary and fruitful, though it was also confusing at times when methodological rigour, positivistic results and also heuristic simplicity suffered at the hands of the sheer variety of approaches.

3 The Main Fields of Emphasis in German Regional Research

In Germany, where this young branch of scholarship initially referred to itself as "spatial research" (*Raumforschung)* rather than "regional research" (*Regionalforschung*), an initial, "politically" oriented point of focus was on a regional science geared chiefly to the governmental regional planning systems, as the state's role in zoning and town-and-country planning has grown, particularly since the end of World War I. Indeed, Erich Dittrich (who played a pivotal role in preparing the famous "vision" chapter in the 1960 "SARO" report, the submission made by the Expert Committee on Regional Planning[16]) believed a quite essential part of the

---

[16] SARO-Gutachten, Die Raumordnung in der Bundesrepublik Deutschland. Report by the Expert Committee on Regional Planning (Sachverständigenausschuss für Raumordnung), Stuttgart 1961.

task of regional research was to develop possible societal visions of spatial planning, and to portray the objective consequences which would arise from each such vision for planning policy.[17]

This orientation towards policy-making was no coincidence. In contrast to the Anglo-Saxon countries, for example, Germany already had some unique precursors of modern regional planning in its Greater Berlin Law (*Grossraum-Berlin-Gesetz*) and the famed Ruhr urbanization management scheme, *Siedlungsverband Ruhrkohlenbezirk* (covering the Ruhr region both in the Rhine Province and the Province of Westphalia) - both of these set up by the new free State of Prussia in 1920. This planning philosophy lived on in the Reich Working Party on Spatial Planning, which later on was quite deliberately transformed into an ideological instrument of government during the Third Reich.

In the event the research activities of German regional science continued to be influenced by this earlier experience after World War II was over, both in the planning activities of the various *Länder* of West Germany and in the framework legislation for federal regional planning which developed rather late in the day; the first draft of the *Raumordnungsgesetz* appeared in 1962, but the law was not finally enacted until 1965. Academics continued to derive both their topical and their more fundamental research issues from the state planning approaches. This discipline has therefore made fruitful contributions to regional planning not just here in Schleswig-Holstein and in other *Länder* but also in developing plans across *Land* boundaries, e.g. in Hamburg and its environs in Schleswig-Holstein and Lower-Saxony. Federal activities did not follow until quite late, culminating in 1969 in calls for a "self-prescribed" federal regional plan, but the concept did not finally take shape until the first *Bundesraumordnungsprogramm* (literally: "federal spatial order programme") in 1975.

At the same time, a critical debate on the growing political and financial interrelations between the federal and *Land* levels in the Federal Republic brought about a fundamental reform in the system of finances for the different levels of state in a unitarian federal state (*unitarischer Bundesstaat*), part of the purpose of which was to compensate for the political inability to redraw the boundaries of the *Länder* themselves, which are far from being equivalent or efficient in terms of size and strength. Thus, a number of "joint tasks" were constitutionally laid down in Articles 91a and 91b of the constitution, as persisting tasks in which federal government bodies would play an accompanying planning and funding role as each *Land* fulfilled its duties jointly with all others. The joint task which particularly attracted scientific research was that of "improving regional economic structure": this was to be examined in terms of the professed policy objectives, its spatial effects, and a number of practices it gave rise to which called for criticism.

[17] Dittrich, Erich, Raumordnung und Leitbild. In: Raumforschung, Bremen 1960, p. 114.

The task called for a sober cost-benefit evaluation in a market economy in terms of the net effects of the combined utilization of the regional planning programme, the fiscal burden-sharing implied and the political impact on the allocation of resources in space, on the regional distribution of incomes and other indicators of the general state of socio-economic development.

In the mean time, regional science in Germany initially gave rather short shrift to another core area of study, namely the development of methodical regional research. In the United States, on the other hand, the trend in the first instance was the opposite way round. However, in the past decades improved access to data and the development of statistical and econometric procedures suited to the characteristics of regional data have had a dominant influence in this country, too, on the application of empirical regional models and the onward development of the methodological principles applied (Walter Buhr, Peter Treuner).

To some extent, regional science here has now simply been catching up with developments in some other countries. In Germany, too, the "quantitative revolution" in economics was a compelling motor for the methodology of regional research. For example, the application of Leontief's input-output approach to the regional scale has come a long way since Isard first popularized it in his *Methods of Regional Analysis*. "Computable general equilibrium" is one of the magic watchwords of today, speaking figuratively too, for a lot of recent progress, especially with complex regional models, has only been made possible by computers.

I do not intend to go into greater detail here on these trends of differentiation and specialization - but this is a sign of a certain degree of maturity in an academic discipline. What I do feel is important, though, is that regional research ought not to be conducted as "art for art's sake", and ought always to be problem-related, interdisciplinary, and action-and-policy-oriented. The Institute of Regional Research in Kiel, as its numerous publications since its foundation bear witness, has always kept its eye on the need to maintain that relevance.

4 Regional Science at the Time of the IfR's Foundation

The growing international significance of regional science as far as the future of industrial societies and the need to cope with their exploded mobility are concerned, a widespread and often intensifying tendency for lands to divide up into conurbations and peripheral regions, but also the potential usefulness of that phenomenon for regional structural policy: these aspects all lent a new prominence to regional science in Germany in the late 1960s and through the 1970s. Initial interest was entirely in a perspective of economies which, although they were gradually opening, were yet still closed *de facto*. Whatever the reason, it was clear even then that the internationalization of industry, the integration of Western Europe to form a common market and the more prominent role of services, science and technology would all need to systematically transcend perspectives as narrow as that - even if this initially only occurred in a familiarization programme of interregional and international comparisons. So this was also a key motive in 1969/70 to found the Institute of Regional Research at the University of Kiel. For quite a number of years, a team of researchers here had been chiefly concentrating on regional research, regional policy and regional planning, who at that time worked in what I had renamed the Department of Economic Policy and Structural Research. Their main topics were infrastructure theory and policy, central locations in rural areas, "joint tasks" for federal and *Land* government, industrialization in rural areas, spatial typology, regional aspects of educational economics, e.g. of universities and their markets, and local government reforms, particularly the redrawing of county (*Kreis*) boundaries in Schleswig-Holstein. A considerable boost came from the German Research Community's newly-formed special research field No. 17 for Scandinavia and the Baltic region, linking the efforts of the universities of Kiel and Hamburg. Specific development-policy projects also deserve a mention, in countries such as Italy, China, Spain, Uganda etc., but particularly also dealing with Schleswig-Holstein itself and its strategic working situation within the common market of "the Six", the divided Germany, and Europe as a whole. So we had the qualified human capital available and the scientific potential, which we sought to get together in the newly-planned institute and hence to channel into systematic, long-term fundamental research.

At the same time, a regional science institute in Kiel was also in the particular political interests of the young *Land* of Schleswig-Holstein, which started to reassert itself as a region in Europe, looking for its specific identity and vision of the future. The *Land* was on the look-out for decision-making assistance, for concepts, instruments and options in its regional planning, administrative structure, regional policy, and much more besides. For at that time, the academic infrastructure available in the field of regional research was largely confined to the southern German universities, particularly Freiburg, Karlsruhe, Tübingen and Munich, together with the University of Münster in the north which established a special research field of "regional research and planning" at the Central Institute of Regional Planning located there. Yet this institution alone was undoubtedly unable to cater for the research interests of the entire region of northern Germany, bordering as it does on Scandinavia and the Baltic region (at a time when Denmark, Norway, the United Kingdom and Eire all appeared set to join the European Community, then still "the Six").

These were arguments which helped the *Land* government and parliament, the university's governing powers (the *Rektor,* the University senate, and also the

Kurator) and others to convince the Schleswig-Holstein Ministry of Education and Culture, and the IfR was able to commence operations in the summer of 1970. It started out in Wik and then in the Dänische Strasse, with two research groups, namely Regional Planning and Educational Economics. The points of emphasis chosen also reflected two urgent issues in the *Land*-political and regional-science settings of the time.

The local government boundary reforms in Schleswig-Holstein, on which we contributed a key report, caused quite a stir. The use of cost-benefit analysis and other new techniques posed significant challenges to town and country planning and regional structural policy in Schleswig-Holstein, as also did the new "joint task" which took effect in 1970 of improving the regional economic structure, the main emphasis of which was on industrialization in rural areas.

This joint task, in combination with the measures to promote the economy of Germany's border regions which had long-since become an institution, took in the whole of Schleswig-Holstein except for the districts of Steinburg and Pinneberg; however, it created a very awkward differential from east to west in the assistance available - instead of from west to east as the region's real structural weaknesses would have required. And of course the IfR's work also included reports with such a key methodological significance yet with rather indeterminate subjects, such as the enquiry into what Secs 3 & 4 of the 1965 Federal Regional Planning Act (*Bundesraumordnungsgesetz*) really meant by the "summary description of all plans and measures with a regional impact".

However, educational economics and its regional impact, too, had its greatest boom during the 1960s. Germany, in fact, was something of a "laggard" among the industrial countries in this respect, as Georg Picht in particular made clear when he spoke of the "educational emergency". The systematic development of the material infrastructure continually lagged behind, whether in education and science, in the health sector, or in energy supply and transport. During the 1960s and '70s, this backlog of problems triggered off a major, frequently controversial discussion as to how much more expansion was needed, and also on the need for reform. The problem was one of "internal reforms", the revaluation of the deutschmark which had now finally become unavoidable, and the prospects for European integration. The peak phase in the actual expansion of the country's infrastructure went hand-in-hand with a boom in infrastructure research. In the field of regional development in its narrower sense, too, the significance of human capital was recognized and instrumentalized at that time.

5 Regional Development with Changing Perspectives

In the course of the last two-and-a-half decades, the Institute of Regional Research has broadened its main research interests. The nucleus up into the 1980s comprised the determinants of regional development and the consequences of longer-term European integration, especially in and around the Scandinavian region. Regional science in Germany also underwent a parallel change in its orientation. With perspectives changing from time to time, the 1970s and '80s saw attention paid at all times to the determinants of, and options for shaping regional development processes, to their close interdependence with external economic influences, but also to their endogenous potential and how that might be strengthened and utilized as the Common Market developed, later to mutate into the Single Internal Market. So here again, the particular phases of dynamic development in this field in Germany tended to follow new regional problem situations and policy tasks as these began to crystallize. After the partition of Germany, and the building of the Berlin Wall in 1961 came - as full employment ended - the moratorium on hiring employees from abroad in 1973, i.e. after a twelve-year period in which the large-scale inward migration from eastern Germany had been deliberately substituted for by millions of *Gastarbeiter* from other countries. This produced significantly changing, different effects. Whereas the East Germans - in a time of full employment - tended to spread out around the country, to settle also in rural areas, the foreigners southerners concentrated in the cities, thus allowing the established industries to continue to operate in their existing locations instead of moving to the rural areas, where the shrinkage of agriculture provided a much smaller, less mobile reserve labour force.

Especially during the 1970s, however, yet another turnabout occurred in the general environment for regional development and regional policy in Germany. From 1973 onwards, there was a dramatic deterioration in the macroeconomic situation. The oil-price shock, compounded by the major revaluation of the D-mark, especially relative to the US dollar, meant that structural change in Germany's industry was tremendously exacerbated, and the hard core of unemployment started to grow and, ultimately, got rigid. The concept of structural weakness and the need for regional-policy and structural-policy action were caught up in an inflationary spiral, becoming "universal" and calling for changed "priorities and posteriorities" - regional and instrumental research and development, technology and innovation, vocational education and training had risen to the status of key concepts. A concommitant effect of this was that infrastructural development moved into the forefront of attention instead of mere investment in plant and equipment.

It was now clear that, in highly industrialized economies well integrated into the world economy like West Germany's, international lines of development also have an increasingly hard-hitting impact on the fate of regions: all of them are affected, though with differences among them, and the consequences have been dramatic since, for the first time, they have hit systematically the stronger, industrialized regions that have started to lag behind in the race to modernize and innovate. The Institute of Regional Research has made its own contribution by producing impressive evidence of these impacts in large-scale empirical surveys of the relative significance of endogenous and exogenous influences upon regional development.[18]

Equally deep-seated effects have come from a dramatic double shift in demographic trends. From about 1975 onwards, the population of West Germany moved into a period of quasi-stagnation. For the first time in this people's history, a phase with a gradually shrinking and steadily ageing population manifested itself without "war and pestilence" taking any responsibility. Yet at the same time, the baby-boom generation brought the prospect of a surfeit, for some years to come, of more people wishing to join the workforce relative to those who were leaving it. This increased the pressures on the educational system and the excess supply on the labour market, while the demand for labour was increasingly directed towards people with better skills and qualifications. In other words, this was a period which brought major challenges, some of them unprecedented in nature, for the supply of infrastructure. These were most pressing in the area of education and training, including both initial training and continual further training for employees, but social and welfare infrastructure was affected too. The issues involved are as urgent as ever in the present day.

It goes without saying that regional science in Germany also took up these challenges in the 1970s, while at the same time new secular trends and qualitative fields came forward to bid for inclusion in the research programme, particularly the complex of environmental problems associated with the way we conduct our lives and our economies. During the 1970s, therefore, it increasingly turned its attention to the negative externalities of urban and regional growth in contrast to a long preceding period in which it had self-evidently assumed that the agglomeration advantages offered by development poles and central locations were a key driving force for development as a whole. A decisive factor in this change in perspective was the observation of two phenomena which had only been recognized in a rudimentary form until that time: one of these was suburbanization and deurbanization; the other was the observation that some large conurbations were now being plagued by above-average, structurally hardening unemployment as they entered into a phase of hysteresis.

[18] E.g. Bröcker, Johannes, Peschel, Karin and Reimers, Walter: Regionales Wachstum und ökonomische Integration, Munich 1983.

Closely associated with this was the emergence of a type of regional problem new to Germany in the 1960s and '70s: that of the monostructural "old-industrialized" areas from the early and high phases of the country's industrialization. This situation virtually begged the shift in the focus and dimensions of regional policy away from the agriculturally dominated rural regions to the large conurbations. At the same time, it called for more up-to-date explanations as to how industrial dynamism and interlinkage relate to other factors, e.g. in the context of product and innovation cycles and their relationship to cycles of urban and regional development. Moreover, the question has had to be newly addressed as to whether governmental bodies have the decision-making competence to actively influence or guide these processes.

As time has progressed, greater attention has been paid to spatial dynamism by regional science. Yet despite these efforts, it is still true today that although a number of long-term driving forces behind regional change have been relatively firmly identified - such as technology, human capital, demographics and the influences of interregional and international linkages - a truly satisfactory explanation (which can therefore be used as an assured instrument of regional development and sustainability) for regional development processes, the role of endogenous potential and the part played by soft locational factors such as cultural facilities and the overall quality of life, has yet to be found. For that reason, the recommendations made to policy-makers are often rather on the scanty side. The main thrust of the advice given is that local decision-makers and societal actors ought to be encouraged and ought to cooperate more closely, instead of pouring more and more financial resources into attracting new businesses from outside a region, that it is important to create an unmistakable profile for a region's development and to nurture a new style of politics which will systematically strengthen the partnership between public and private sectors, and that regional management should look to generate and utilize synergy effects from networking and cooperating on a partnership basis with other regions in Europe and around the world. These can hardly be said to be a new patent medicine, but they are pragmatic, in some cases new approaches which it is up to each region to make the most of in its own specifically creative way.

6 Current Challenges to Regional Science

Since the beginning of the 1990s there has been yet another fundamental change - which had been totally unpredictable - in the environment in which regions develop, bringing a dramatic shift backwards to geopolitical patterns of some time ago. The caesura which has occurred with the unification of the two German states, the reopening of the central and eastern European economic region and the prospect of economic and monetary union in a steadily growing European Economic Area all pose new challenges to regional research in Germany and elsewhere in Europe.

In this altered environment for regional development, significantly enough, older research themes in regional science are enjoying a renaissance. For example, when it comes to the task of bringing on the new *Länder* of former East Germany so that they can attain the economic standards of western Germany as swiftly as possible, the infrastructural prerequisites of structural change and economic development deserve particularly close attention. However, research interest has also grown immensely in the European dimensions of infrastructure provision, partly due to growth considerations, as this is seen as a vital prerequisite for developing the economic opportunities provided by the Single Market. However, there is also a regional-policy aspect in that the common or rather single market is a precondition for attaining relatively harmoniously paced growth throughout Europe as a whole.

It is important here to overcome the process in which agglomeration zones flourish with regional disparities growing, which is a result of market integration and is encouraged by integration policy, or is at least favoured by it. And peripheral regions find themselves excluded, more or less: these are trends and tendencies which are not effectively resolved by action financed by the European Structural or Cohesion Funds. Such agglomeration zones include the western European "blue banana" which was the key north-south development axis, and which left aside the peripheral regions in the West, the South and the North, not to speak of the effects of the former Iron Curtain to the East.

Now, after 1989, the west-east development dimensions in the larger European region cry out for revitalization and more efficient linkage, exacerbating the northern and central parts of the "blue banana" slightly more than the southern ones. These agglomeration areas are contrasted by the peripheral regions which do not automatically stand to gain from a highly integrated economic zone. It is worth noting that large-scale outward migration, "passive rehabilitation" and rigorous pay differentials along productivity lines have so far been more or less ruled out de facto by the European Union's agenda of policy objectives. All this makes it imperative that the call for an active, autonomous regional structural policy in the various individual regions must take hold. In this context, of course, it is not merely an academic question as to how to strengthen the position of Schleswig-Holstein, as a "bridge" or a subject between central Europe, Scandinavia and the Baltic countries.

Within this field of research activity there is a need for the analysis of the regional conditions for and impacts of integration. And of course that reactivates some of the issues which the Institute of Regional Research in particular has focused upon during the last two decades, namely: What consequences will these newly delimited environments have for the international and inter-regional division of labour? Will the differentials in regional levels of development around Europe be exacerbated or diminished as the integration process marches on?

The old controversy as to whether European integration has tended to encourage spatial concentration in the central regions of western Europe and will continue to do so in future, or whether integration helps to even out regional divergences has still not been settled, even after all of the enquiries made into the issue, especially in connection with the then-impending Single Market and now newly impending European Monetary Union,[19] be it formed in one step or at different speeds.

7 European Economic and Monetary Union

As things appear today, it is virtually inevitable that a multi-level strategy of integration will have to be followed in the interests of achieving progress in integration, but there are many risks involved in this. Any development of a schism between "first" and "second" divisions needs to be avoided or at least minimized. If, as is quite probable, montary union proceeds on a principle that "to be beautiful it will have to be small", something will have to be agreed as to what practical ways can be used to leave the door open for other countries to enter later on. If it is not, too high a price will have been paid in terms of integration policy.

Precautionary measures also need to be taken on the currency front. To avoid the risk of the various parts of Europe successively drifting apart, we need to determine what form of currency-policy arrangements should be made to maintain a link between the countries not yet participating in the final stage and the first group. This cannot be done on a freely chosen, *à la carte* basis with countries choosing their arrangements in isolation. At all events, we shall continue to need a viable, renewed, if only looser European Monetary System to interlink the monetary union with the other member states which have not yet joined.

Otherwise the ensuing uncertainties about the real terms of trade might put the Single Internal Market in danger of being split. Whether because weaker-currency countries indulge in competitive devaluation or because they relax their discipline with regard to qualifying for entry to the monetary union as soon as possible, the probabilities of regional disparities will be growing. What is needed is that the

---

[19] E.g. Jochimsen, Reimut: Perspektiven der Europäischen Wirtschafts- und Währungsunion, Cologne 1994.

disciplinary influence on the behaviour of its member countries, stipulated by the Maastricht Treaty, in their monetary, exchange-rate, fiscal and wages policies, must further dampen down inflationary potential in all countries, to cure budgetary imbalances on a sustained basis, and hence to encourage the swift introduction of the common currency in the other member states too.

All these considerations apply in the same way to future member countries. If they do not, I repeat, this will generate a possible threat to the Single Market (and the European Economic Area) which at present certainly do not yet form an "optimum currency area". For the hard-currency countries the risk would be that inflation is imported, and for the reforming countries it would be a disfigurement of the real economy in terms of growth, employment and structural renewal. So it is ultimately in their own interests that the prospective new members fulfil the minimum preconditions for participation in the EMS.

To this end, a certain minimum level of price stability must already have been attained, the central bank must be independent, it must have the power to control the supply of money, and must be forbidden from lending directly to government bodies; additional requirements are currency convertibility and free payment-clearing and capital flows. Underlying such a "nominal alignment" it is essential for the real economies to become more closely interlinked, among the effects of which would be to bring cyclical movements and response patterns closely into line. That would make exchange-rate adjustments as a result of so-called "exogenous" or "external shocks" less and less necessary.

Once the third stage of EMU is reached, it will be up to the member states, which primarily means up to the governments in their budgetary policies and up to other economic actors, to adhere strictly to the "rules of the play" within the monetary union. In collective bargaining too, therefore, the room for manoeuvre in distributing resources will be restrained more strictly and more acutely than is already the case in an open market by the relative progress in improving productivity in particular regions and sectors. This will be all the more keenly felt since the last resort of adjusting exchange rates to cope with particularly sharp differentials in economic and socio-structural development will now have been relinquished once and for all. A new trend is becoming clear for the eventual participating countries - especially if they have had a relatively weak currency and structurally weak regions in the past - that regional policy will increasingly also have to play the part of an exchange-rate policy. Any such regional structural policy, especially for the benefit of regions which are structurally weak or having to cope with industrial restructuring, calls for adequate financial resources and transfer payments which pose an additional burden on the balance of resources within each state. That soon creates a call for EU funds, thus reactivating Germany's experiences since 1990, which are still far from over and done with. That in turn raises issues of efficiency, monitoring and cost-benefit analysis both for specific

specific measures and the instruments of regional structural policy: these are issues which have frequently been raised by regional science.

8 Deepening and Widening the EU

In the mean time, regional researchers, like the makers of regional structural policy, now have another essential dimension to consider. It is important today to consider a new quality of the major European economic zone now building up, which must include not only the old EC of "the Twelve" but also Scandinavia, the Alpine countries, the southern and eastern Mediterranean shores as well as the central and eastern European countries whose economies are catching up: in the long run, most of these countries must participate on an equal basis. There can be no questioning the fact that, as a continental economy forms on this scale, it will be impossible either to have an executive based centrally in Brussels or to have national governments capable of ensuring that growth, employment and the quality of life are effectively managed. The governmental proposals for the Maastricht-II conference tend in any case to reflect hopes of renationalization rather than a will to adopt a longer-term perspective of progressive political integration with the ultimate objective of a democratic, federally-structured, social body-politic. In the face of economic globalization and the resulting employment crisis, the issue also arises as to what form of European constitutional structure can provide optimum economic development, a growth and employment dynamic with a rational spatial allocation of resources and a high level of integration among the economies as well as the regions they encompass. Against this background, it will be up to regional science to make its own contribution towards ensuring that opportunities to strengthen and make fuller use of endogenous potential in the regions, while also exploiting their potential for cooperation within wider area concepts of development impetus, are examined afresh and are seized whenever they can be.

As I have suggested,[20] such strengthening of the regions is particularly sorely needed because the European Single Market whose impact will bite more and more strongly, the proposed support of that market by European monetary union from 1999 onwards, and the extension of the European Union to take in anything up to 27 member states will not only give added impetus to the integration of markets, but at the same time will lend a thoroughly new quality to the dissolution of boundaries within Europe - a process which has been under way throughout the post-war era. This is a phenomenon, as it were, of a double dissolution of the nation states' and national economies' boundaries, which could only be redressed or

[20] E.g. Jochimsen, Reimut: Perspektiven der ..., op. cit., p. 152.

cancelled out by making tremendous economic, employment, social and political sacrifices. Added to the economic dissolution of boundaries brought about by market-driven Europeanization and globalization, there is a parallel trend in the political sphere (i.e. European Integration) as nation states have been renouncing autarky and protectionism, and will continue to have to renounce the powers traditionally available to them in the fields of monetary, exchange-rate, fiscal and employment as well as redistributive policy.

However, the dissolution of boundaries in Europe also simultaneously means a fading in the effectiveness of the traditional governmental tools used to compensate for adverse developments or for underdevelopment in the regions, particularly as those undergoing structural change are now exposed to ever tougher and more extensive competition from other locations. Many industrial locations in the west, which had the protection of the "Iron Curtain" until 1989, now have their low-paying competition not in eastern and southern Asia as before, but in the reforming countries of central and eastern Europe, which in some cases makes them just a one-day round-trip away by lorry. And the diminishing of the relative costs of transport, even when corrected by ecological externalities, will continue. Simultaneously, the internationalization of production and factor procurement is increasingly also now embracing not just labour-intensive branches of manufacturing but also the services which are becoming more and more tradable on a supra-regional basis. Even vital areas of infrastructure for large geographical areas - such as telecommunications and energy supplies - now face this opening-up process and are to become objects of competition.

In the face of these challenges, the question of what avenues of action are now open to regional policy-makers needs to be placed right at the top of regional science's research agenda. This not only embraces location theory, but also growth and socio-economic development theories. Gunnar Myrdal, Harvey Leibenstein, Albert O. Hirschman with their "vicious" and "virtuous circles", their "backwash and spread-effects" as well as their "forward and backward linkages" will now challenge the evolving pattern of integration of highly developed economies and regions. The question is: To what extent will structurally weak regions in the European mega-economy be forced in future simply to lay themselves open to "passive rehabilitation", and to what extent can an active structural policy counteract this process of desiccation by "raising" the endogenous development potential with the help of technology policy and regional technology transfer, occupational training and labour-market structural policies? What opportunities are offered by cooperation and by measures to ensure fair competitive conditions?[21]

[21] Cf. the experience with such policy effort in North-Rhine-Westphalia since the early 1980s. See Jochimsen, Reimut: Technologiepolitik für die 90er Jahre, Konzeption und Umsetzung am Beispiel Nordrhein-Westfalens, in: Bentele, Karlheinz, Reissert, Bernd

The fundamental question spanning this entire situation is therefore what integration concept and what distribution of tasks and responsibilities are needed and are the right ones for the European Union. Will it, for example, be sufficient to have a free trade zone with a common currency, or do we not really need a combined package of economic, monetary and political union? I firmly believe that the latter is the case. Any examination of the history of monetary unions will also show that the constitutional establishment of a regional system for EMU is a vital issue. The new advisory "Council of the Regions", established under the Maastricht Treaty, does of course offer the beginnings of a more systematic involvement of the regions in formulating Union policy, but is that really enough? Is it not necessary for the regions' codeterminative influence in the European Union to be substantially enhanced, also to truly contribute to a vision of a "Europe of the regions"? Exploiting national differences as well as linguistic and cultural diversity, the wealth which these regions may represent and contribute provide essential prerequisites for economic efficiency, social stability, ecological initiative as well as democratic vitality and identity.

9 New Visions

We can be quite certain that the structural challenges facing the regions as Europe grows together will be a focal point of public discussion and, I trust, of research in German regional science right up to the turn of the century and beyond. Yet this alone still provides an incomplete picture of the present-day agenda for regional science. For, in parallel to these developments, the ecological challenge is giving rise to totally new needs for visions of regional development, which in fact are increasingly transcending national boundaries. The first steps in this direction have already been taken with the "Basic Principles of European Regional Development Policy" and with the common vision up to 2010 prepared by the countries round the shores of the Baltic Sea. Regional research is now called upon to play its own part in designing a viable concept for ecologically as well as economically sustainable regional development. This, in other words, has to be an approach which will help achieve a breakthrough for solutions which technologically integrate the right kind of care for geographical space, and the resources associated with it, into production processes, products themselves, transportation, housing, and all parts of our economic structure and living conditions. Only then can we be truly considerate in our use of energy and resources while simultaneously establishing a sustainable basis for industrial competitiveness and strengthening it in the long run. This new vision can only be implemented on a collaborative basis across

---

and Schettkat, Ronald (eds.): Die Reformfähigkeit von Industriegesellschaften, Lectures to the honour of Fritz W. Scharpf, Freiburg i.Br. 1995, pp. 199-221.

national borders. I am speaking of collaboration among political bodies, with an internationalism of the kind which the academic world has long-since put into practice - as demonstrated by the 25-year history of the IfR. My best wishes to you for the next 25 years!

Economic Integration and the Space Economy: Lessons from New Theory

Johannes Bröcker
Technische Universität Dresden
Institut für Wirtschaft & Verkehr
D - 01062 Dresden

1 Introduction

Soon after the IfR was founded, Karin Peschel in 1973 initiated a research programme aiming at analyzing the spatial impact of economic integration. As usual, at the beginning of this research project stood a screening of existing theories. Peschel presented the results from her first screening to the European RSA congress in summer 1974. They were published as no. 10 in the institute's discussion paper series (Peschel 1974).

She first discussed the Heckscher-Ohlin theory, concluding that, according to this theory, one should expect regional inequalities to diminish in the course of integration. She was critical about the validity of this proposition, however, for two reasons. First, regional comparative advantages could hardly be regarded as exogenous, at least in highly developed parts of Europe. Second, integration effects stemming from decreasing cost in a larger market were neglected in the theory.

As an alternative approach she discussed the potential model, routed in regional science traditions (Clark et al. 1969). According to the potential model, footloose industries were predicted to move closer to the center of gravity of the integrated Europe. It was therefore suspected that integration might aggravate rather than mitigate regional inequality. Among people concerned with European regional policy this was a widespread conviction at that time. It motivated the expansion of the means for regional policy in the structural fund of the EC.

Though a plausible story could be told why high potential areas in the center of Europe might attract footloose industries, other stories also sound plausible. The theory behind the potential approach remained obscure.

The conclusion from this screening of theories was that a firm theoretical basis for predicting spatial effects of economic integration was out of sight. Peschel's consequence was to pursue an eclectic approach, to state a series of hypotheses and check them against available evidence. You know that this strategy was successful. Clearly, the research programme would have failed, if she had postponed the empirical work until a more satisfying theory was available.

My lecture is about progress in theory since that time. I talk about growth theory and location theory. Both have gone through a fundamental revision during the last decade. I will briefly introduce the main ideas and then explain some implications for the spatial effects of economic integration. On the basis of the new location theory I also draw some conclusions about European regional policies.

2 Growth Theory

2.1 What is the Novelty of New Growth Theory?

New growth theory started in the ending eighties with the work of Paul Romer (1986, 1990) and became a huge field of research all over the world, with fundamental contributions by the world's leading economists like Lucas, Barro, Grossman, and others. The advanced text books by Grossman and Helpman (1991) and by Barro and Sala-i-Martin (1995) show the state of the art.

A new growth theory was called for, because the old theory based on the work of Robert Solow (1956) was unable to explain long run growth and to give advice how governments could influence it in a desirable way. Solow's theory could only explain the transition to a long run steady state, but the long run steady state growth itself is exogenous in the theory. Technical progress rains down from heaven and makes labour steadily more and more productive. Capital accumulation has to follow the path prescribed by exogenous technical progress.

Economists were, of course, unsatisfied with this exogeneity of growth in this theory, ever since Solow introduced it. But any attempt to make long run growth rates endogenous was condemned to failure, as long as the conventional neoclassical basis of the theory remained untouched. Endogenous growth and conventional neoclassical theory are incompatible. In the following under the term "conventional neoclassical theory" I subsume all kinds of perfect competition models with price taking rational decision makers, without externalities and with convex technologies. Remember that convex technologies exclude economies of scale.

In conventional neoclassical theory the following "impossibility of growth" theorem holds: If some essential production factors cannot be augmented, and if technology is constant over time, then there is no long term growth. I call a production factor "essential", if it earns a strictly positive income share.

Solow's way out was to dispense with the constant technology assumption and to introduce exogenous technical progress instead. New growth theory wants to avoid this easy way out, and therefore has to depart from conventional neoclassical theory. The stock of knowledge is introduced as an additional factor of production. Investments into the stock of knowledge have to be produced using resources withdrawn from the production of goods. In this framework long term growth is possible, if augmentable production factors like knowledge and capital have jointly non-decreasing returns to scale in the aggregate production function (i.e. if their respective partial production elasticities add up to an amount not less than unity). This implies that the entire production function has increasing returns to scale. This is so, because the production function contains, beyond the augmentable factors just mentioned, non-augmentable production factors like labour and land, which have strictly positive partial production elasticities. Hence, the elasticities of all factors add up to an amount strictly larger than unity.

The theoretical difficulty is how to reconcile these increasing returns with a consistent theory of markets and factor remuneration.

Different solutions have been suggested in the literature:

(1) Knowledge as an externality (Romer 1986, based on Arrow 1962 and Sheshinski 1967): This approach takes knowledge accumulation as a by-product of production or capital accumulation. People learn by doing. Producers can use the stock of knowledge as a free good. Marginal productivity theory can then still be applied to all other factors, which have jointly constant returns to scale. The problem with this approach is that private firms have no incentive to produce knowledge, because it is free. Hence, the uncomfortable pure externality assumption has to be made.

(2) Knowledge produced by the state (Barro 1990): Here the state collects taxes for producing knowledge, which, as in (1), is then offered for free.

(3) Private knowledge production: In this approach knowledge comes in the form of patents for new goods (Romer 1990) or new methods (Aghion and Howitt 1992). Firms have a monopoly on patented goods and remunerate knowledge producers by monopoly profits. Labour and capital get less than their marginal products due to monopolistic price mark-ups.

The third approach is close to the story told by Schumpeter (1912) in non-formal terms. It is the most interesting one and the most widely used now. Working through the technicalities of this approach is all but simple, but the main messages can be intuitively understood without going into the formal structure.

I have commented on the general policy implications of new growth theory elsewhere (Bröcker 1994). Thus here I will confine on the implications for the integration issue.

2.2 New Growth Theory and Dynamic Integration Effects

The new theory opens the door to a rigorous theory of dynamic integration effects. According to the theory integration boosts growth. The mechanism is as follows: If more people join the integration club, the cost of a new idea per club member declines because of non-rivalry in the use of knowledge. Hence, the aggregate efficiency of knowledge production increases. As a consequence, the growth rate gets larger, because knowledge production is the engine of growth (Rivera-Batiz and Romer 1991). The details of the mechanism are more complicated, of course. One has to check, among other things, that integration does not induce resources to be withdrawn from knowledge production to an extent outweighing the efficiency gain. Grossman and Helpman have shown how such adverse effects may take place (Grossman and Helpman 1991, Ch.9). The main message is the one explained, however.

In the real world, however, the integration area is not just a homogeneous club which can be smaller or larger. If a spatial dimension is introduced, we have to be aware of the regions' differing starting positions and differing locations. In a many-countries or many-regions world it is not guaranteed that each region equally participates in the dynamic gains. It is not even guaranteed that each region gains at all. The very same assumption making growth possible, namely non-decreasing returns of augmentable factors, also implies that diverging growth is possible.

Whether or not growth in an integrated economy of several regions actually diverges depends on the assumptions on factor mobility and knowledge diffusion. If the effect of knowledge accumulation on growth is purely local, and if there is some mobility of non-augmentable factors, then growth must diverge (Grossman and Helpman 1993, p. 25). The other extreme is purely global effects of knowledge accumulation. This means that each new idea has the same effect on aggregate output at each location. Then convergence is obtained, if some constant factor is immobile. The mechanism generating convergence is the same as in Solow's model under these conditions.

The more interesting world of complex spatial structures of growth is to be found between these two extremes. What all can happen in such a world with limited mobility of factors and knowledge is not yet sufficiently well understood to allow simple conclusions. Uwe Walz from Mannheim worked through the complexities of this world in a series of papers (Walz 1995a, 1995b). Even though fairly special assumptions (e.g. zero transport costs for final goods) have to be made to keep models manageable, the arithmetic is toilsome. The taste remaining after reading it is a bit like that you have after reading books on modern industrial economics:

Depending on the details, everything can happen. Hence, we are back where Karin Peschel was already two decades ago: Let the data speak.

3 Location Theory

3.1 Old Location Theory

A general equilibrium theory of spatial economies already existed before the new location theory came up in the beginning nineties. The seminal work of this kind is, of course, von Thünen's (von Thünen 1826), the founding god of spatial economics, as Samuelson (1983) has called him. Von Thünen's model, venerable as it is, meets all standards of conventional neoclassical theory. The monocentric structure of "Der isolirte Staat" is very special. But we may take this simplicity as a pure pedagogical device, and any kind of multi-centric pattern of market places can be introduced. Debreu (1959) has already noted that equilibrium in a space economy is included in general equilibrium theory as a special case. It is just a matter of interpreting goods, preferences and technologies.

But just like the sources of growth are exogenous in Solovian growth theory, the sources of location are exogenous in von Thünen-type location theory. Von Thünen's central market place is given by assumption, not explained. Spatial variations in factor endowments in Heckscher-Ohlin type theories are given by assumption, not explained. Alonso's CBD is exogenously given, not explained. Just like Solovian growth theory explains the allocation of factors over time induced by an exogenous engine of growth, von Thünen-type theory explains the allocation of factors in space following an exogenously determined economic geography.

The stagnation theorem showed that it's not by coincidence that sources of growth have been taken as exogenous in the old growth theory. The exogeneity of location is no coincidence in conventional neoclassical location theory either. One way to prove this proposition is Starrett's spatial impossibility theorem (Starrett 1978). It says that in a conventional neoclassical space economy with homogeneous space no spatial division of labour can exist if transport is costly. Hence, a theory explaining spatial structure endogenously must depart from the assumptions of conventional neoclassical theory.

August Lösch (1943) was obviously well aware of that. Though he explicitly wanted to generalize Walras' theory, his assumptions were non-Walrasian. Economies of scale and monopolistic competition were the corner stones of his theory of an economic landscape. But unfortunately, Lösch was only partly successful. He suggested a consistent model for a single homogeneous goods market, but his attempt to arrive at a general theory by composing partial equilibria, fascinating and intuitively convincing as it is, is full of contradictions and inconsistencies. Go through his marvellous book and try to consider how it might be formalized with clear assumptions about technologies, preferences and market structures, with budget constraints and equilibrium conditions. Be sure: You get hopelessly lost.

This made regional economists believe that any attempt to set up a consistent theory of endogenous location was condemned to failure.

3.2 New Location Theory

Help, then, came from another side than expected. The break through of a new location theory was Krugman's book on geography and trade (Krugman 1991), though others like Rivera-Batiz (1988) already came close to his ideas before. Until its publication even regional scientists carefully avoided the word geography. They were afraid that the audience, expecting slide shows of industrial landscapes, would fall asleep. Now the well known Krugman not only talks about geography, he even puts the word in the title of his book!

What made him and other leading persons in the profession think about economic geography. Definitely not the tradition of location theory. Krugman is not very friendly with it. He writes in his preface: "Although the intellectual tradition of location theory is both wide and deep, what is taught is usually a very narrow set of geometric tricks involving triangles and hexagons." (p. x)

Like any new theory, this one was triggered by external and internal forces. The internal force was that coming to location theory was just the logical next step in the new trade theory, which had developed during the eighties, and to which Krugman made main contributions (Helpman and Krugman 1985). One of the starting points of the new trade theory was to explain trade between countries not differing significantly with respect to endowments. Hence the theory had to give an endogenous explanation of regional specialization patterns. From there it was only a small step to introduce a true spatial dimension by allowing for positive transport costs.

The external force triggering new location theory is world-wide integration and a growing interest in the spatial impact economic integration. Integration increases the interest in the space economy simply because countries are going to be no interesting entities anymore. Countries become regions in a larger market. Not their belonging to a national state, but their position in space is going to be their important characteristic.

Which question aims the new theory to answer? What is its methodology, and what are the answers?

Imagine a homogeneous space. Let the same technology be available all over space, and let transport be costly. How then can we explain the emergence of cities, of densely and sparsely populated areas, of spatial division of labour, and of trade by pure economic forces? Obviously, the theory must describe the interaction of those forces, which the late Edwin von Böventer (1979) has called the "raumdifferenzierende Faktoren": Economies of agglomeration exerting a centripetal force, the necessity to use land in production and consumption, exerting a centrifugal force, and transport cost making it advantageous for different activities to be close to each other.

This, of course, has always been the programme of endogenous location theory. What makes the difference is the methodology. The new theory seeks to meet all standards of modern theory of general equilibrium. The working of the economy should be well defined by assumptions about preferences, technologies and market structures. Economic behavior should be derived from optimization of rational decision makers, there should be a well defined equilibrium concept, and all budget constraints should hold.

Meeting these standards is not without costs. One has to dispense with the full generality of the Arrow-Debreu world. General equilibrium modelling with imperfect competition and economies of scale is only possible if models are tiny and special. The number of regions, factors, and sectors is small (in the order of one or two). Preferences and technologies are represented by tractable parametric functions like CES or simpler ones. Let's have a look at an example.

Figure 1 shows the "tableau économique" of a 2-regions economy, which has been studied by Krugman and Venables (1994). Typically there is a constant returns sector producing a homogeneous good under perfect competition (call it agriculture) and a sector producing a set of diversified goods under monopolistic competition (call it manufacturing). Manufacturing firms are vertically linked, because the manufacturing output bundle is also an input to manufacturing. Both goods are tradable. Trade in manufactures is costly.

There is at least one immobile factor preventing all activity to become concentrated in a single location. A mobile factor can be added to make the model more realistic, but the essentials would not change.

The monopolistic industry is the source of agglomeration economies in the system. There is a large number of firms, each producing a different product variety under increasing returns. Each firm sets its price with a monopolistic mark-up on marginal costs. Free entry into the market drives profits down to zero.

An ingenious trick invented by Dixit and Stiglitz (1977) allows for a simple formalization of the monopolistic industry. The trick implies that all output variations in the industry are variations in the number of firms, while output per firm remains constant. It is assumed now that customers prefer product diversity. A more diversified industry makes downstream firms more productive and consumers more happy. This is why market size is a factor of attractivity in this class of models, contrasting with conventional neoclassical models, where market size is simply irrelevant. By the Dixit-Stiglitz trick economies of scale on the level of a firm are translated into economies of agglomeration on a regional level.

Figure 1 "Tableau économique" of an endogenous location model

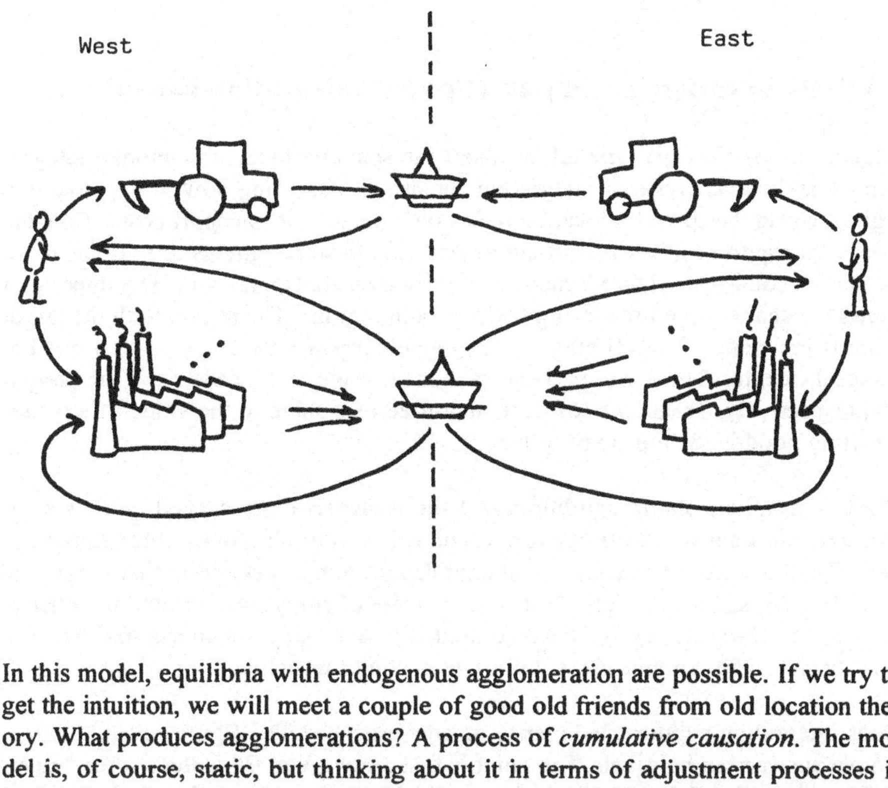

In this model, equilibria with endogenous agglomeration are possible. If we try to get the intuition, we will meet a couple of good old friends from old location theory. What produces agglomerations? A process of *cumulative causation*. The model is, of course, static, but thinking about it in terms of adjustment processes is easier. Start from a symmetrical allocation, and move some manufacturing firms from West to East. Then two disequilibrating forces begin to work, making East even more attractive. The first force are *backward linkages*. Firms moving eastwards increase demand in East, making it more attractive for other firms to locate in East. The second force are *forward linkages*. More manufacturing in East allows Eastern firms to buy the same product diversity for lower prices due to lower transport costs. In regional science language they gain from a higher supply potential. Manufacturing firms follow manufacturing firms. This is the centripetal force.

The countervailing centrifugal force is exerted by the immobile factor. Workers are required for production, and they are a market for manufactures.

When a model of this kind is solved for a system with many regions or even with continuous space, complex economic landscapes arise. Recently, Tomoya Mori from Pennsylvania managed to solve a continuous space model with several sectors on a computer (Fujita, Krugman and Mori 1994). The outcome is an interesting complex economic landscape, very much reminding of a Löschian central place system. Unlike Lösch's system, however, this solution is perfectly consistent in the sense of general equilibrium theory.

3.3 New Location Theory and Spatial Integration Effects

What do models of this type tell us about the spatial effects of economic integration? A stylized analysis of integration consists in observing how in a simple two regions world the spatial allocation reacts on a decline of transport costs. Take the words "transport cost" as a short-cut of any kind of impediments to trade between regions or countries. Figure 2 maps equilibria over transport costs. The upper part shows the share of manufacturing firms in both regions. The region with the larger share is the center, by definition. Due to symmetry it is arbitrary, which one becomes the center. The lower part shows the real wage in the center and the periphery, respectively. Transport costs are measured as a share in the value of a manufacturing good in the region of origin.

There is a unique stable equilibrium, if the transport costs exceed t_1. It is symmetrical. For transport costs between t_0 and t_1 five equilibria exist, three stable and two unstable ones. The symmetrical equilibrium remains stable in this range, but there are also stable equilibria with a large share of firms concentrated in either of the regions. If transport costs become smaller than t_0, only the concentrated equilibria survive. The symmetrical equilibrium becomes unstable.

Does integration support divergence or convergence? The figure reveals that no simple answer can be given. The point is that among von Böventer's three "raumdifferenzierenden Faktoren" two play a trivial role: Agglomeration economies favour concentration, immobile resources draw into the opposite direction. But what about transportation cost? Are they an agglomerating or deglomerating force? Students like to have a simple answer, but there is no such answer.

High transport costs force economic activity to locate close to immobile resources. Medium ones allow firms to gain from locating close to each other. Clearly, this is only possible if prices for immobile factors are higher in the agglomeration than in the periphery. This exerts a deglomerating force. If transport costs decline even further, this deglomerating force must eventually dominate. This is so, because gains from agglomeration are also based on transport costs. If transport costs vanish, these gains vanish either. Where we stand in this figure is an empirical question. It is exactly the question to which empirical research in this Institute was directed over more than a decade.

Figure 2 Equilibria over transport costs

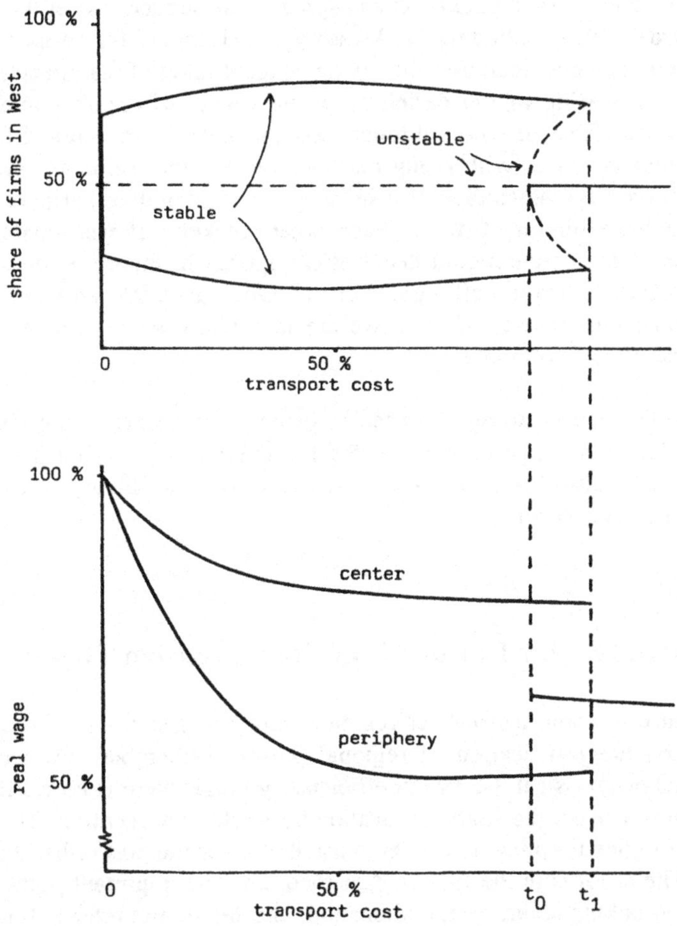

What about welfare in the two regions? The lower part of figure 2 gives the answer by mapping real wages in both regions over transport costs. There is no income other than wage income in the model. Hence, the real wage is the appropriate measure of utility. Real wages attained under zero transport costs are set to 100%. In the symmetrical equilibrium prevailing in disintegrated world, utility is, of course, the same in both regions. It increases due to increasing gains from trade, if transport costs decrease. As long as the symmetrical equilibrium remains stable, both regions equally share in integration gains. At a certain critical point t0, however, the system switches to an asymmetrical equilibrium. The center jumps to higher, the periphery to lower welfare. With transport costs decreasing further, the periphery eventually recovers again.

Trying to find an intuitive interpretation of this process, again two good old friends come along, called backwash and spread. The sudden loss of the periphery is what Myrdal (1959) called the backwash effect. Demand for peripheral factors of production suddenly decreases due to the concentration of manufacturing in the center. The recovering of the periphery in the course of ongoing integration is Myrdal's spread effect. It works through two channels. First, demand for the peripheral output increases with rising real income. Second, firms are increasingly willing to exploit the advantage of a lower price of immobile peripheral factors, because the force binding firms to each other weakens. If we depart from the assumptions of the conventional neoclassical approach, theory alone cannot tell whether integration favours all regions or not. Obviously, it is an empirical question of utmost importance, whether we are in a phase where backwash or one where spread effects dominate.

In my opinion there is strong empirical evidence that we are in the latter phase, the spread-dominance phase, as far as the EU is concerned. Other papers of this conference will raise this issue. My contribution is about theory, and I will not pursue this here any further.

3.4 Regional Policy Lessons from New Location Theory

Instead, I turn to some lessons for regional policy in general. I will confine my comments on the justification of regional policy. Remember that according to conventional neoclassical theory, on efficiency grounds there is no justification of any attempt to correct the spatial allocation by market intervention. The only justification of regional policy, if any, is to use it as a second best redistributional instrument. The proof is in the first welfare theorem. The important point, however, is this: When talking about spatial allocation, one should not refer to this theorem, because its assumptions are necessarily violated in an economy with an endogenous economic landscape. This is an implication of Starrett's theorem!

The first welfare theorem is not applicable to market equilibria of the new theory. And actually they are all inefficient. There is always room for Pareto-improving the spatial allocation brought about by the market. Whether this calls for state intervention, however, is another matter. The answer is all but obvious. When we discuss market interventions, the welfare cost of market failure has to be weighed against the welfare cost of state failure.

I know that some of you are not keen scholars of von Hayek. But whatsoever, one has to accept that the state has an information problem. A necessary condition for the state's ability to Pareto-improve the allocation is that it can get the information about how to bring that about. There must be simple rules to be applied.

If we observe regional policy in practice, actually we find some general rules and might ask, whether their application is likely to enhance efficiency. Take the regional policy of the EU as an example. To a large extent it consists in gathering money in the center and pumping it to the periphery. This is a rule simple enough to be applicable even for a state which is badly informed as compared to the mass of private decision makers. But is it a good rule on efficiency grounds? The new theory gives no indication for that.

In standard models of the new theory concentration is not generally too strong. If anything, it is too weak. Concentration may fail to occur at all, though it is efficient. It may also happen, that concentration occurs, but even more concentration is required for efficiency. It may furthermore happen that activities agglomerate in a wrong place due to historical coincidence, or that the number of cities in the economy is inefficient. Inefficient spatial patterns survive due to coordination failure. It may well be preferable, if all firms or households jointly change location, but actually they don't because individually they better stay, where all others are located. This is the well known network externality, which may lock in the economy in an inefficient state. To summarize, many kinds of inefficiencies may occur. The specific pattern of agglomeration may be wrong, but there is nothing wrong with agglomeration in itself. If there is any obvious misallocation, it is too little rather than too much concentration.

Pointing to the problems of congestion and pollution in cities is no valid objection against this proposition. Agglomeration occurs even though people are well aware of these problems. Higher land prices in the center are a sufficient proof, that the economies more than outweigh the diseconomies. For small decision makers these economies are external. This is a strong indication that markets spontaneously produce too little rather than too much agglomeration (Matsuyama and Takahashi 1993). It is interesting that, on a small regional scale, regional policy accepts this proposition, concentrating investments in central places. But on a larger, on a national or European scale, an equalizing policy is pursued.

It is not impossible to design a model in the style of the new theory, with a market generating more concentration than desired. That means, self-defeating regional concentration, as Matsuyama and Takahashi (1993) have called it, is principally possible. But the story told by a model with this property sounds not very plausible empirically.

The question whether the new theory makes a case for regional policy for the sake of equity is a different matter. The new theory clearly undermines the confidence in the conventional neoclassical proposition that welfare levels tend to converge in an integrated economy. The results show that in the course of integration peripheral regions may lose, not only in relative but even in absolute terms! Therefore a compensation may be required for reasons of fairness.

To summarize: The case for equalizing regional policy got stronger on equity, but not on efficiency grounds. I obtained the same conclusion in my discussion of regional policy lessons from new growth theory already referred to above (Bröcker 1994).

4 Final Remarks

What did we learn from the new theories? Did we learn anything? Critics of rigorous model building deny. "We knew all these things before!" they are shouting. "Cumulative causation, forward and backward linkages, economies of agglomeration working against the deglomerating force of immobile resources, spread and backwash, ancient stories all that. A few guys from MIT are half a century late in beginning to understand the space economy. After having read a textbook they translate it into equations and claim to have invented a new theory!"

Well, this is the "it's all in Smith"-argument. It is always true and therefore never helpful. I am convinced that in economic theory we will never see new results which have not been known before somehow. Theoretical progress mostly has the form that things which we were talking about in a bad language are translated into a better one. A better language is not necessarily mathematics, and putting things into equations does by no means say that the language has improved. A better language is a language imposing a discipline on the way of reasoning. It forces us to deduce conclusions logically from precisely stated assumptions. The new theories discussed are progress in this sense. They are also informative for policies, allowing new insights about valid and non-valid justifications of regional policy.

To what extent does the new theory inform empirical research? Here I am a bit more critical. Let's make a thought experiment: Imagine a restart of Peschel's empirical research programme about economic integration and the space economy. Would we ask very different questions, and would we apply very different methods in the light of the new theories? I doubt. More or less the same things would have to be done with the only difference that now they are done with the blessing from main-stream theoretical economics.

References

Aghion P. and P. Howitt (1992), A model of growth through creative destruction, *Econometrica*, 60:323-351.

Arrow J. K. (1962), The economic implications of learning by doing, *Review of Economic Studies*, 29:155-173.

Barro R. J. (1990), Government spending in a simple model of endogenous growth, *Journal of Political Economy*, 98, part II:S103-S125.

Barro R. J. and X. Sala-i-Martin (1995), *Economic Growth,* McGraw-Hill, New York.

von Böventer E. (1979), *Standortentscheidungen und Raumstruktur*, Schroedel, Hannover.

Bröcker J. (1994), Die Lehren der neuen Wachstumstheorie für die Raumentwicklung und die Regionalpolitik, In U. Blien, H. Herrmannn und M. Koller, Hrsg., *Regionalentwicklung und regionale Arbeitsmarktpolitik*, Beiträge zur Arbeitsmarkt- und Berufsforschung, Bd. 184, S. 29-50, IAB, Nürnberg.

Clark C., F. Wilson, and J. Bradley (1969), Industrial location and economic potential in western europe, *Regional Studies*, 3:197-212.

Debreu G. (1959), *Theory of Value*, Wiley, New York.

Dixit A. K. and J. E. Stiglitz (1977), Monopolistic competition and optimum product diversity, *American Economic Review*, 67:297-308.

Fujita M., P. Krugman, and T. Mori (1994), On the evolution of hierarchical urban systems, Paper presented at the European Congress of the Regional Science Association in Groningen.

Grossman G. M. and E. Helpman (1993), Endogenous innovation in the theory of growth, NBER Working Paper No. 4527, Cambridge, MA.

Grossman G. M. and E. Helpman (1991), *Innovation and Growth in the Global Economy.* MIT Press, Cambrige, MA.

Helpman E. and P. R. Krugman (1985), *Market Structure and Foreign Trade: Increasing Returns, Imperfect Competition, and the International Economy*, MIT-Press, Cambridge, MA.

Krugman P. (1991), *Geography and Trade*, MIT Press and Leuven University Press, London.

Krugman P. and A. J. Venables (1994), Globalization and the inequality of nations, CEPR Discussion Paper No. 1015, London.

Lösch A. (1943), *Die räumliche Ordnung der Wirtschaft*, Gustav Fischer, Stuttgart, 3. Auflage, 1962 (1. Auflage 1943).

Matsuyama K. and T. Takahashi (1993), Self-defeating regional concentration, NBER Working Paper No. 4484, Cambridge, MA.

Myrdal G. (1959), *Ökonomische Theorie und unterentwickelte Regionen*, Gustav Fischer, Stuttgart.

Peschel K. (1974), Integration and Regional Economic Growth, Institut für Regionalforschung, Discussion Paper No. 10, Kiel.

Rivera-Batiz F. L. (1988), Increasing returns, monopolistic competition, and agglomeration economies in consumption and production, *Regional Science and Urban Economics*, 18:125-153.

Rivera-Batiz F. L. and P. M. Romer (1991), Economic integration and endogenous growth. *Quaterly Journal of Economics*, 106:531-556.

Romer P. M. (1986), Increasing returns and long run growth, *Journal of Political Economy*, 94:1002-1037.

Romer P. M. (1990), Endogenous technological change, *Journal of Political Economy*, 98:S71-S102.

Samuelson P. A. (1983), Thünen at two hundred, *Journal of Economic Literature*, 21:1468-1488.

Schumpeter J. (1912), *Theorie der wirtschaftlichen Entwicklung*, Duncker & Humblot, München, 2. Auflage, 1926 (1. Auflage 1912).

Sheshinski E. (1967), Optimal accumulation with learning by doing, In K. Shell, editor, *Essays on the Theory of Optimal Economic Growth*, pp. 31-52. MIT Press, Cambridge, MA.

Solow R. M. (1956), A contribution to the theory of economic growth, *Quarterly Journal of Economics*, 70:65-94.

Starrett D. (1978), Market allocations of location choice in a model with free mobility, *Journal of Economic Theory*, 17:21-37.

von Thünen J. H. (1826), *Der isolirte Staat in Beziehung auf Landwirtschaft und National-ökonomie*, Friedrich Perthes, Hamburg.

U. Walz (1995a), Does an enlargement of a common market stimulate growth and convergence? Institut für Volkswirtschaftslehre und Statistik der Universität Mannheim, Beiträge zur angewandten Wirtschaftsforschung, Discussion Paper 524-95, Mannheim.

U. Walz (1995b), A dynamic model of economic geography and the enlargement of a common market, Institut für Volkswirtschaftslehre und Statistik der Universität Mannheim, Beiträge zur angewandten Wirtschaftsforschung, Discussion Paper 525-95, Mannheim.

Testing Models, Describing Reality or Neither? Convergence and Divergence of Regional Growth Rates in Europe during the 1980s[1]

Paul Cheshire and Gianni Carbonaro*
London School of Economics and Political Science
Department of Geography
Houghton Street
London WC2A 2AE

1 Introduction

There has recently been increasing interest in regional economics. Indeed Blanchard (1991) goes so far as to claim that "macroeconomists have rediscovered regional economics". Krugman has coined the phrase "the new economic geography" (Krugman, 1991). One recent manifestation of this interest has been the lively debate about 'convergence' and 'divergence' in growth rates (Baumol, 1986; Barro, 1991; Barro and Sala-i-Martin, 1991; Barro and Sala-i-Martin, 1992; Sala-i-Martin, 1994). To regional economists this is not a new debate. It can be traced back at least to Borts and Stein (1964) and the debate between the followers of Myrdal (1957) and Kaldor (1970) and the neo-classical school. This debate was summarised by Richardson (1973), and more recently by McCombie (1989a and b), and has formed a staple of most regional economics courses ever since (see, for example, Armstrong and Taylor, 1993).

It is argued in this paper that the claim which is implicitly made - that testing for, and finding 'convergence' in regional or national per capita incomes constitutes a test of neo-classical growth theory - is methodologically flawed; both in principle

[1] The authors would like to acknowledge the support of the Economic and Social Research Council under grant No. W113251003 and Lend Lease International. The authors would also like to acknowledge the help and criticism of colleagues, while equally absolving them from any responsibility for remaining errors. Special thanks are due to Bruno Giussani who was the senior research fellow on the project for Lend Lease International. Collecting and analysing the data for European urban regions for the 1991 Census round involved much research assistance and thanks are particularly due to Chris Brooks, Jack Cheshire and Rachel Kerrison.

* European Investment Bank, Luxembourg

and in measurement terms. The most important reason for scepticism with respect to the actual measurement of convergence is the arbitrary definition of the set of 'shocks' which have been included in such tests. One of the shortcomings of the set of variables additional to the 'convergence' parameter which have been included in the models of regional convergence has been the neglect of spatial factors influencing regional growth rates. Using a new data set for comparably defined European city regions, it is shown that apparent convergence or apparent divergence can be 'found' depending on how the model is specified. The most complete specification generates a result which, in the terms of the current debate, suggests there is neither convergence nor divergence over the time period considered.

2 The Convergence/Divergence Debate

The renewed interest of macroeconomists in regional economics, at least as manifested by Barro and Sala-i-Martin (1991), is in the neo-classical tradition. It presents itself as a test of what is essentially a Solow (1956) steady state model of growth. The growth equation derived by Barro and Sala-i-Martin (1990, 1991) was formulated as a log-linear approximation from the transition path of the neo-classical model for closed economies. The model takes the following form:

$$(1/T)\cdot \log(y_{it}/y_{i,t-T}) = x_i^* + \log(\hat{y}_i^*/\hat{y}_{i,t-T}\cdot\left(1-e^{-\beta T}\right)/T + u_{it} \qquad (1)$$

where:

i	indexes the economy,
t	indexes time,
y_{it}	is per capita output,
x_i^*	is the rate of technical progress,
$\hat{y}_i^*$	is the steady state level of output per effective worker,
T	is the interval between observations,
ß	is the convergence coefficient to be estimated, and
u_{it}	is an error term.

The model actually estimated is:

$$1/T\cdot \log(y_{it}/y_{i,t-1}) = \alpha - \left[\log(y_{i,t-T})\right](1-e^{-\beta T})(1/T) + \text{other variables} + \varepsilon \qquad (2)$$

where:

ε	is a stochastic error term.

The source of convergence in the model is the central assumption of diminishing returns to capital. The specific mechanism in which this is embodied is a Cobb-Douglas production function although - so long as the production function embodies diminishing returns to capital - other forms would appear possible.

The speed of convergence is measured by the β parameter and will be increased if labour mobility is allowed in the model, since movement of labour from the labourich, poorer region to the richer region will boost returns to labour in the poorer and accelerate the onset of diminishing returns to capital in the richer region. A similar increase in the rate of convergence would result from technological flows from the richer to the poorer region. It should be noted that the model is set up and calibrated in such a way that the estimated value of β measures the annualised 'rate of convergence' and a positive sign means that poorer regions grow faster i.e. there is convergence.

The model (1) includes an error term, u. This, however, is not an ordinary error term but is composed of two elements: a truly random term plus a set of 'shocks' which affect different economies in systematically different ways. Since new 'shocks' may increase the actually observed dispersion of (regional) incomes over time a positive β does not necessarily imply a falling dispersion of observed (regional) incomes. Indeed Barro and Sala-i-Martin (1991) distinguish two forms of convergence: so-called β convergence and σ convergence, the latter referring to the standard deviation of the log of incomes at any time. β convergence, they claim will provide an answer to the question how long is it likely to be before the per capita income of a particular country catches up to the average per capita income across economies, *ceteris paribus;* σ convergence tells us how the distribution of observed per capita incomes has changed over some particular time period.

3 Test or Measure?

Barro and Sala-i-Martin (1991) then go on to apply this model to data for US states for the period 1880-1988 and to a subset of European regions for the period 1950-85. This work has created much interest and lively debate. It has already been shown that their measure of convergence is sensitive to their choice of European regions and the definition of data. Armstrong (1994) concludes that using essentially their methods but improving the data and including more regions from low income countries of the European Union (EU) greatly reduces the estimated rate of β convergence. Quah (1993) has shown that the underlying implicit assumption - that there is an identifiable and consistent growth rate in a given economy - is not consistent with the observable facts and that even if the measured rate of β convergence is positive this does not necessarily imply a falling dispersion of per capita incomes. The underlying position adopted in this paper is, however, more radical. It is argued that measured β convergence neither constitutes a useful test of theory nor does it provide a robust measure of convergence/divergence.

The reason why it is believed that it does not constitute a test of growth theory is that it cannot distinguish between possible explanations. A finding of positive convergence is not inconsistent with the predictions of steady state growth models but it is an indirect test and also consistent with other possible explanations. It is therefore almost the weakest test possible. It would be much more persuasive if the mechanisms generating convergence within the model - diminishing returns to capital or the income equalising effects of labour and capital mobility - were directly verified. There are various possible mechanisms that might generate convergence; diminishing returns to capital is only one such mechanism. Labour mobility is another. More damagingly, however, any observed convergence could simply reflect the impact of policy. Achieving regional convergence has, after all, been a significant goal of national policy in most of the countries of the European Union (EU) throughout the period analysed by Barro and Sala-i-Martin (1991); it has been a central goal of EU policy since the mid 1970s. There are still other possible mechanisms that could produce convergence in regional incomes. Thus, even if we are confident we have measured convergence accurately (and the evidence of this chapter suggests we cannot be) the result is only not inconsistent with neo-classical growth theory.

It is perfectly possible - perhaps more realistic (see below) - to take the view that there are mechanisms tending to generate convergence and others, such as increasing returns (Kaldor, 1970; Krugman, 1991), generating divergence. In calibrating the model the estimated value of β is therefore only a measure of their net effect over the time period in question. In other words to interpret the results as a test of the neo-classical steady state growth model you have to exclude other possible explanations *ab initio.*

It is with the second criticism, however, that this paper is chiefly concerned. As was discussed above, Barro and Sala-i-Martin (1991), and other researchers using this framework, divided the error term in the model into two components: a random term and another to represent 'shocks' having a systematic effect on (regional) economies. Although the term 'shocks' is used, effectively this second element of the model should capture any systematic effects influencing regional growth rates since those are the dependent variable. If it does not, the estimate of the β coefficient will be biased and non-robust. The variables used to capture these 'shocks' have typically been very simple measures of the economic structure of activity in the region (although this implicitly introduces the complexity of non-homogeneous output) and dummies to represent the countries in cross regional/country analyses.

The problem is that the set of potential 'shocks' which should be allowed for to get unbiased estimates of β cannot be specified *a priori*, so that if there are omitted variables which are systematically correlated with growth rates it is not clear what value to attach to claims of convergence or divergence. This situation is made worse because of the cavalier approach that has been adopted to the choice of variables to capture the effect of shocks. The choice seems narrow and arbitrary and inevitably excludes any measures of ways in which regional economists would expect spatial effects to operate.

4 The Present Approach

This paper turns this procedure on its head. The aim is first to define regions in an economically coherent way. It will be argued that regions, cities and states are not interchangeable observational units as implied by Blanchard (1991) (who claims that the reason why macroeconomists have rediscovered regional economics is that "regions offer them a new, larger data set; there are many more regions [cities, states]..."). It is important to have consistent definitions. The second aim is to devise variables which reflect systematic spatial influences on the differential growth of European city regions. The third aim is to formulate variables in ways which simultaneously allow us to examine the plausibility of the assumptions underlying European regional policy. The fourth aim is to try and specify the influences on the growth performance of city regions as fully as possible, including a spatial adaptation of Romer's (1990) endogenous growth model. This last, it is argued, will - *ceteris paribus* - tend to produce divergence of regional growth rates. Finally, because of the underlying scepticism with respect to the interpretation of the β coefficient, the model is set up in the simplest possible way and simply includes the log of the initial per capita income level to test for convergence or divergence.

As was noted earlier there is an active policy in the EU designed to generate regional convergence. The European Commission has had three principal policy concerns. Structural policy has targeted aid to the backward agricultural regions of southern Europe and to declining industrial regions; it has had a long-standing concern for the problems of 'peripheral' regions (in contrast to the 'core' regions) - which the process of European integration is thought to make worse; and, most recently (Millan, 1993), there has been concern for the '... diversification of disparities... generating a patchwork in which privileged areas border directly on depressed areas'. The empirical work reported here attempts directly to test some of these policy concerns. Is the diagnosis of regional growth processes implicit in European regional policy concerns, consistent with the data? Given spatial adjustment mechanisms, how can we reconcile increasing mobility within Europe with an increasing 'patchwork'? Does the process of European integration tend to assist core regions at the expense of peripheral ones?

In the US the assumption tends to be made that interregional labour mobility is so strong that characteristic specific income or unemployment differentials are equalised on the margin.[2] In Europe both policy makers and academics tend to assume that barriers to mobility mean that interregional equalisation will be far from the norm. Even in a neo-classical perspective 'equilibrium differentials' are the expected outcome (Evans and Richardson, 1981, Gordon, 1985). At an intraregional level, however, positions are broadly reversed. Europeans, analysing compact city-regions which, socio-economically and ethnically are, by US standards, comparatively homogeneous, argue for equalisation on the margin (Cheshire, 1979, Cheshire, 1993). Americans, however, express continuing concern for implicit disequilibrium at the intraregional level (Wilson, 1987, Hughes, 1993). The argument in this paper is that although both the US and European views of spatial adjustment involve strong assumptions, each view is more appropriate to its continental context. There are very significant barriers to interregional migration in Europe but intraregional adjustment to demand shocks is speedy and complete because of induced adjustments to commuting patterns. In the context of densely urbanised regions, such as the Randstad, Ruhr, or the English Midlands or South East, this mechanism of spatial adjustment extends to interregional adjustment, helping systematically to generate Millan's 'patchwork of privileged areas (bordering) directly on depressed areas'.

5 The Definition of Region

The focus in this paper is growth in regional per capita income.[3] Regions, however, are not defined as administrative units[4] but as Functional Urban Regions (FURs). FURs are quite close in concept to the SMSAs used by Glaeser et al

[2] The work by Glaeser *et al* (1992) explicitly assumes that in the US migration equalises marginal real wages to skill specific labour across US city-regions and the implicit assumption of such a strong definition of spatial equilibrium is common to much US analysis (e.g. Blomquist *et al*, 1988; Gyourko and Tracy, 1991). A comparable assumption is usually made with respect to the interregional equalisation of characteristic specific unemployment differentials (Marston, 1985).

[3] Converted at OECD purchasing power parities into $US. This eliminates the effects of relative but not absolute price changes.

[4] The N.U.T.S. (Nomenclature des Unités Territoriales Statistiques) of EU data and policy- in Germany, for example, the Länder or Kreise. In the US the nearest equivalent would be counties, states and regions.

(1992) but they are more generously bounded. They are defined in terms of concentrations of employment. To each concentration of employment - or core - are added all spatial units from which more workers commute to the core in question than to some other core. This set of spatial units forms the hinterland of each core, so that each FUR consists of a city-core and its labour market sphere of influence.[5]

For reasons explained at much greater length in Cheshire and Hay (1989), regions so defined are economically coherent, being as relatively self-contained as spatial subdivisions of a single open economy can be. Apart from being desirable in itself this has the specific advantage in the present context of abstracting from the distorting effect of commuting flows. The richest regions in the Barro and Sala-i-Martin data set included three that are 'underbounded' metropolitan regions: Paris, Hamburg and Milan. Because there is net inward commuting into these regions income per resident is inflated. Between 1950 and 1985 inward commuting into this type of region grew significantly, artificially inflating their measured growth performance. Since the boundaries of FURs include their commuter fields, using them as the oberservational units minimises this problem, although as is shown below, it does not eliminate it.

Another advantage of using FURs as our observational units is that they are representative measures of local economies (not, for example, excluding poor industrial zones) and they are appropriate for policy analysis since - because of their relative self-containment - the effects of policy interventions within them are likely to be largely contained; in contrast to interventions at a more local level which will spill over into surrounding areas (Gordon and Lamont, 1982).

No official data are published for FURs except in the UK (and then on boundaries not identical to those used here). Demographic and some other data have been collected from national sources for the smallest administrative units of each country and aggregated to the FUR boundaries originally defined by Hall and Hay (1980). From these, weights can be derived to apply to official Eurostat data for the smallest (Level 3) European administrative regions. Where two FURs lie wholly or mainly within one Level 3 region, national data have been used to generate estimates for each FUR separately. This allows reasonably accurate estimates of comparably defined FUR data, such as GDP or unemployment rates, to be generated. For more detail as to the methods used see Cheshire and Hay, 1989. The data

[5] FURs are typically more self-contained than Travel to Work Areas (which have fixed levels of self-containment) but the precise degree of self-containment will vary according to how closely they are clustered. Cities such as Madrid, Toulouse and Plymouth have very extensive hinterlands and are highly self-contained because they do not compete with other employment centres. Cities such as Essen, Coventry or Bruxelles, close to other large cities, are less self-contained.

analysed here relate only to the European Union's (EU) largest FURs, those with a core city of 200.000 or more and a total population of a third of a million or more. There are 122 such urban regions in the EU but four, those of Greece and Portugal, have had to be excluded because of lack of data.

6 The Model

Having entered the log of GDP per capita, the next variables to be entered are designed to capture national influences on regional growth. In the convergence literature this is typically done simply by entering a national dummy. This is an unnecessarily crude approach. FURs are not isolated from the rest of the economies of which they are a part. National macroeconomic policy, education and training policies, national culture, legal and social institutions and, most importantly, the temporal incidence of the economic cycle, all influence measured rates of economic growth of both national economies and of their component city regions. These are entered as the growth in per capita income in the national remainder (that is, excluding the FURs that are the object of the analysis) over the 1979/82 to 1987/90 period. The largest city regions analysed here account for approximately half of total economic activity.

This has the advantage of entirely avoiding any definitional correlation while entirely capturing the policy, cultural and institutional context (except in so far as those systematically differ between large city regions and the rest of their nations). It implies that the national context in which city regions are located is a cross-sectional constant influence on their growth rates rather than being nationally specific as a national dummy implies.

There are, however, some identifiable and additional national factors that need to be allowed for. The first of these is the knowledge that German cities were influenced by the influx of refugees from Eastern and Central Europe, notably from the former DDR, in the period under study (only FURs of the former West Germany are included in this analysis for reasons of data availability). It is well documented that these flows were substantial (see, for example, Burda (1993)). Other work (Cheshire, 1995) has shown that German FURs displayed substantial population growth in their cores over the 1980s as a whole, reversing a decade or more of population loss. Although this was part of a wider pattern of partial urban population recentralisation in northern European city regions during the 1980s, elsewhere there was mainly relative, rather than absolute, recentralisation. The explanation seems likely to be the influx of refugees/migrants in 1989 and 1990. Just as this raised population so it would have tended to lower per capita income growth, other things being equal, because the new migrants remained mainly unemployed.

The German border with Belgium is virtually open and other evidence suggests that lower living costs in Belgium have induced a movement of population working in north west Germany to live in Belgium. Independently of this there has been extensive immigration to Brussels. The four major city-regions of Belgium were therefore included with those of Germany in this dummy variable.

The final variable reflecting national factors is a dummy for Northern Italian FURs (including Rome). The rate of growth of the national remainders acts as a proxy for national, institutional, social and cultural factors but the difference between northern and southern Italy in these respects is exceptional. This dummy is to allow for the possibility of two Italies, therefore.

The next set of variables reflect the definition of 'shocks' as used be Barro and Sala-i-Martin but are formulated in ways which try to reflect EU regional policy assumptions. The first of these is designed to capture any systematic spatial effects of European integration. The Single European Market and earlier steps towards economic integration have been presented as being to the benefit of all European citizens. In reality, however, the impact must create a potential Pareto gain rather than a pure one since the mechanisms producing the gain are expected to have differential industrial impacts and involve the exploitation of comparative advantage and economies of scale. In addition, since the elimination of invisible barriers to trade is analytically equivalent to a reduction of transport costs, if there are any economies of scale available to be exploited then the results of Krugman (1991) would suggest these would be spatially concentrated.

Empirical evidence for the spatial influence of economic integration in Europe is relatively scarce but a fact of political bargaining and of policy making in Europe is that steps towards integration are believed to strengthen 'core' regions at the expense of 'peripheral' regions. That is why each step towards further integration since 1970 has been accompanied by a strengthening of structural policies to reduce 'spatial disparities'. If only for policy reasons, therefore, it is important to test for the impact of European integration on regional growth differentials.

The concept of economic potential (defined as the accessibility to total income at any location, allowing for distance, transport costs and tariffs) as put forward by Clark *et al* (1969) is an *ad hoc* one in theoretical terms. Various theoretical reasons why it might be associated with regional growth differentials and justifications for the use of changes in it as a measure of the spatial impact of European integration are possible (Cheshire, 1995b). It remains a fact that there is no alternative measure of the systematic effects of integration. Since the focus of the present analysis is growth, the appropriate indicator is change in economic potential brought about by tariff reductions, falling transport costs and EU enlargements. This is calculated from Clark *et al* (1969) (who considered the regions of a Europe of 10, including Norway but excluding Greece) supplemented with estimates derived from Keeble *et al* (Keeble et al, 1988) for the regions of Spain. These have been scaled to Clark's values. The resulting values are shown on Map 1.

Map 1: Change in economic potential as a result of European economic integration and falling transport costs

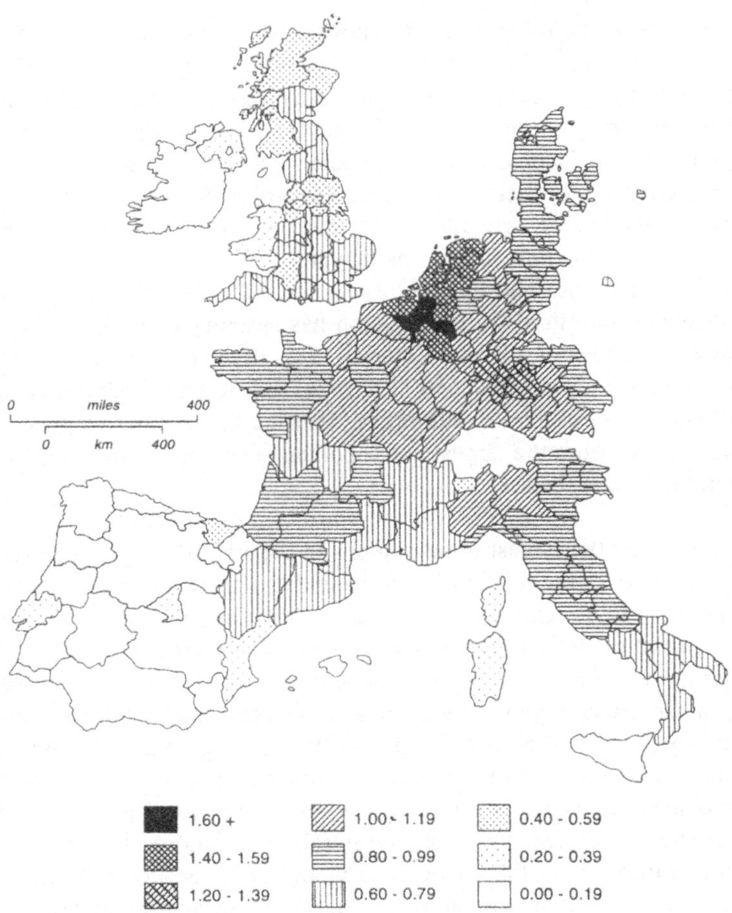

Source: Clark et al. (1969); values for the regions of Spain and Portugal estimated on comparable basis from Keeble et al. (1988).

The theoretical arguments as to why integration should favour core regions do not imply that that relationship measured for the 1980s should be linear. Clark's calculations are for different hypothetical states of the world but with data for 1966. Induced growth may have been fastest where economic potential increased most in the initial stages of integration and falling transport costs. But this growth would tend to bid up land and labour costs and produce additional congestion. In turn this would tend, over time, to produce deconcentration from the core to surrounding regions, so by the 1980s the relationship in terms of Clark *et al* (1969) estimates should be quadratic.

Variables reflecting two other types of 'shocks' have been formulated in a way which reflects European policy concerns. European regional policy identifies two main categories of region for assistance; the so-called Objective 1 regions are the backward agricultural regions, mainly of southern Europe. That Objective 1 regions have per capita income levels well below the EU average is self-evident since that is a criterion for their designation. Not self-evident, however, is whether a heavy representation of agriculture and low income is associated with slow growth of per capita income. This is particularly pertinent given the rapid trend towards industrial decentralisation observed since the 1970s. There is a long standing view that industrial decentralisation has systematically been directed to more rural regions (Fothergill and Gudgin, 1982; Keeble *et al*, 1983). The regional growth model of von Böventer (1975), indeed, stresses both intra and inter urban agglomeration economies and implies that the fastest rates of growth will, through time, move from the largest, metropolitan centres to successively smaller and more distant urban centres.

In the context of all the largest urban regions in the EU the most appropriate expectation should be for the relationship between FUR growth in per capita income and specialisation in agriculture in the wider surrounding region to be quadratic. FURs in the most urbanised regions would have slower growth but - because of decentralisation to relatively close and more rural areas - more agricultural employment in the wider region would initially be associated with faster growth. This effect would tend to fall away and in FURs surrounded by very agricultural regions - the backward peasant agriculture of much of southern Europe - growth in per capita incomes would be slower. They are too remote from advanced metropolitan centres to receive significant decentralisation and rural-urban migration of unskilled workers from the countryside would be associated with population increasing faster than output and falling average levels of human capital.

The European Regional Development Fund (ERDF) also identifies so-called Objective 2 regions - old industrial regions experiencing industrial decline and employment loss - as suffering from problems.[6] Not all industries are declining - par-

[6] These are identified in 1988 on the basis of three criteria : i) that the unemployment rate in the three years prior to 1988 exceeded the Community mean; ii) in some year(s) since 1975 the proportion of industrial employment has exceeded the Community mean; and

ticularly in value added terms - and data permitting the systematic identification of particular industries, even for the largest administrative units, N.U.T.S Level 1 regions, are not easily obtained,[7] much less so for FURs. In addition it is reasonable to argue that the influence of some industries, especially coal, continues adversely to affect local growth prospects for a considerable period after they have ceased to account for a substantial share of local employment and output. In the case of coal, other industries which have developed because of its local availability tend to decline or disappear, land has to be reclaimed if it is to be redeveloped, the environment has been damaged, local skills are inappropriate and perceived industrial relations problems may act as a disincentive for inward investment. By 1979, the first date for which the data on GDP per capita be calculated, the coal industry had entirely disappeared from Belgium and the Netherlands and all but disappeared from north eastern France and the Ruhr. The historic influence of the coal industry is measured in this model, therefore, by a dummy variable related to the physical coincidence with the area of FURs; if the whole of a FUR was located within the area of a coalfield (as defined in the *Oxford Regional Economic Atlas,* 1971) it had a value of 2; if a part was within a coalfield area it had a value of 1; and otherwise it had a value of 0.

The local effects of the port industry have had much less attention than many traditional declining industries such as coal, steel or shipbuilding, but the dynamic effects of the transformation of the port industry on local economies may well have been more important. The introduction of containerisation and roll-on roll-off ferries did not just result in a loss of employment and a reduction of value added in port activity but it meant that quite rapidly port cities ceased to exercise a locational attraction for processing industries[8] and even to some extent, for distribution. Again the historic influence of port activity would seem to be the appropriate variable. It is measured as a dummy variable scaled from 0 to 4 strictly according to the value of seaborne port traffic in 1966 (*Oxford Regional Economic Atlas,* 1971). The largest inland port on the Rhine is also included.

These two variables - the influence of the coal and port industries - might be seen as proxies for traditional resource based industrial specialisation which, in as far as it was associated with human capital, was associated with very industry-specific

---

that iii) there had been an observable fall in industrial employment in the region compared to the year(s) selected for ii).

[7] It might be thought that employment by sector would be readily available but different national classification systems and simple non-availabality (in Italy before 1981 for example) mean that unless very large numbers of observations are dropped, only a three sector classification - agriculture, industry and services - can be calculated for any date before about 1986.

[8] See Alonso (1964) for a discussion of the historic locational attractions of ports for processing industries in their capacity as transshipment points.

skills. A final industrial structure variable is more generalised. The 1975 percentage of the labour force engaged in industry in the wider Level 2 region in which the FUR is located is included to reflect the susceptibility of the urban economy to shocks to the industrial sector as a whole. It should be noted that there is a very low correlation between this variable and the percentage in agriculture in the same year.

The variables discussed so far all relate to European policy concerns. There is considerably more theoretical interest in the role of specifically spatial economic factors since these have been entirely ignored in the literature. The first two are relatively straightforward. Different activities tend to concentrate in cities of different sizes, as the size at which diseconomies of agglomeration offset the economies of agglomeration varies between types of activity. Those activities which were fastest growing in the 1980s, such as business services, finance and information intensive activities, tend to be concentrated in larger cities.[9] Thus we might expect city size to be positively associated with income growth but, given the evidence on the distribution of city sizes, this would be more appropriately entered as the log of population. After standardising, so far as possible, for other factors influencing the rate of income growth, we might also expect a measure of density to be negatively associated with growth. There are two possible but not mutually exclusive reasons for this. The first is that higher density, other things being equal, is likely to be associated with greater unit congestion and infrastructure costs; the second is that it is likely to be a reasonable proxy for land rents. The higher these are, other things being equal, the lower income growth (both absolute and per capita) is likely to be.[10] Density is calculated for the area of the FUR as a whole in 1981.

The next variable reflects the spatial adjustment process. As was noted above net commuting flows affect the calculation of regional GDP per capita and changes in them will affect measured growth rates. As a generalisation, in the US it is typically assumed (in the non-convergence literature) that real wages for characteristic specific (that is skill, experience, etc.) labour equalise on the margin between city-regions (Marston, 1985; Blomquist *et al* 1988; Gyourko and Tracy, 1991; Glaeser *et al*, 1992). In Europe, with its much higher costs associated with migration, this

[9] It has been widely suggested that small and medium sized cities grew fastest during the 1970s and 1980s (though this has usually not been tested in a multivariate context and it is not clear whether the implied dependent variable is total population, total income or per capita income). However the size of cities in the present analysis has been necessarily truncated; only large FURs are included because of the high cost of collecting and adjusting data collected by national agencies. In 1981 the FURs which are the object of the present analysis accounted for just over 50 per cent of EU population.

[10] This is not to deny that other factors, such as natural or policy created restrictions on urban land supply, will also influence land prices.

is not a reasonable assumption. Equilibrium differentials are likely to remain even in marginal returns (Gordon, 1985) and greater ones are possible with respect to average returns (Cheshire, 1993). As was pointed out in Cheshire (1979), however, two forms of spatial adjustment process exist in the labour market; migration, involving a change of both workplace and residence - or discontinuous mobility - and alterations to commuting patterns - or continuous mobility. Since the potential for continuous spatial mobility exists every time an individual changes jobs, and the flow of job changes is very large (see Bowers *et al*, 1972), continuous mobility represents a low cost adjustment mechanism in densely urbanised regions.[11] Thus the more densely FURs are clustered the more induced commuting changes will be produced for a given change in income differentials and the more likely it will be that characteristic specific incomes (or unemployment probabilities) will equalise on the margin.

The effect of this mechanism will, however, be paradoxical with respect to measured per capita income growth. There will be two effects. The first will be that the FUR in which incomes per worker grow faster will attract commuters from its neighbouring FUR(s)[12] thus increasing its workforce but not its measured resident population. Thus measured per capita income will grow even faster than income per worker as a purely statistical artefact. There is likely, however, to be a real effect too. There is a strong positive relationship between length of commuting trip and income and measures of human capital. There is also such a relationship between area of search and measures of human capital. Thus the FUR in which per capita incomes grow faster will tend systematically to attract more productive and highly paid workers from its neighbouring FUR(s) increasing the differential in mean per capita incomes further by a form of composition effect.

We should therefore expect that, other things being equal, per capita income will grow faster, in a faster growing FUR, the nearer it is to other, less rapidly growing FUR(s); and that conversely in the FURs where per capita income grows more slowly, it will grow more slowly still the closer the FUR is to other FUR(s) where income is growing more rapidly. It should be expected that there will be some lag in this process since commuting patterns do not adjust instantaneously so the variable is measured over the period 1979-86. This also avoids any possible problem of definitional correlation. Differences in per capita income growth were meas-

---

[11] It should be noted that this is a fact of the European settlement pattern and therefore of its spatial adjustment mechanisms. City regions in the US tend to be much more isolated, so adjustment via induced commuting pattern changes between (but not within) city regions is unlikely to be very significant except in some specific areas.

[12] Thus extending its boundaries if these could be perfectly and continuously measured but the units analysed in this paper have boundaries defined with respect to 1971 data and 'frozen'.

ured for all FURs within 80 kms of each other and this differential was divided by distance to reflect the expected distance decay effect.[13]

The seminal contribution of Romer (1990) was to formulate a consistent model of economic growth which makes technological progress endogenous and respond to market incentives in a believable way. A key result is that since there are increasing returns to human capital employed in research, because of the non-rival and partially non-excludable nature of a part of the output of knowledge generation (the 'blueprints' or designs), larger markets can induce more knowledge generation and so not just create static increasing returns to scale but dynamic increasing returns to scale in the form of a faster growth rate. The mechanism which produces this result is the rate of growth of A, the number of new designs, ideas or blueprints. This is a function of human capital employed in research. Thus in a Romer-style model the quantity of human capital available, and the proportion devoted to research, drives economic growth. This derives from his equation (3):

$$\dot{A} = \delta H_A A \tag{3}$$

where: $\dot{A}$ = the rate of growth of non-rival knowledge
H_A = total human capital employed
A = stock of knowledge in research.

Although he acknowledges other assumptions are possible he assumes "anyone engaged on research has free access to the entire stock of knowledge". A more plausible assumption to an urban economist, used to thinking about the source of agglomeration economies, industrial districts and the like, is that there is both a distance decay function associated with access to knowledge[14] and agglomeration economies in new knowledge production. The more opportunities there are of bumping into someone working on a related problem or with useful, incremental knowledge, the more knowledge productive the individual worker is likely to be.

---

[13] Some FURs within 80 kms of each other are divided by a national boundary. Later experiment showed that in these cases there was no measurable effect whatsoever; but the variable on which the results reported here were estimated was calculated doubling linear distance if a national boundary intervened.

[14] Universities are themselves embodiments of this phenomenon. Access to the knowledge of colleagues is strongly associated with the number to whom one has easy access: the nature of coffee room discussion is knowledge spillovers. In the present case I would like heartily to acknowledge the contribution of a graduate student - Stefano Magrini - with whom I have enjoyed many hours discussing ways of applying Romer-type models in a spatial context.

This idea can easily be incorporated within a modified version of Romer's equation (3)

$$\dot{A}_r = \delta H_{Ar} \cdot A \cdot \phi_r \tag{4}$$

and

$$\phi = H_{Ar}^{\alpha} \tag{5}$$

where:
- r denotes the region
- ϕ denotes knowledge spillovers
- α a parameter describing the concentration of human capital within a region and the resulting opportunities for productive contacts.

This reduces to

$$\frac{\dot{A}_r}{A} = \delta H_{Ar}^{1+\alpha} \tag{6}$$

A reasonable argument may be made that this will capture both differential spillover effects of knowledge within regions and between regions. The chance of accessing useful knowledge available in another region will increase as your opportunities of meeting someone who has recently visited this region, or is networked into that region, rise. Lone researchers have the same probability of accessing knowledge wherever they are (the Romer assumption) but the more concentrated they are, the greater their probability of accessing knowledge both existing within the region and in other regions. This suggests that there will be dynamic increasing returns to the spatial concentration of human capital via accelerated knowledge generation. H_{Ar} will be raised to a power of $1 + \alpha$ and $\alpha > 0$.[15]

---

[15] This is strictly only an explanation of disequilibrium growth differentials since it would imply that all research and development would ultimately end up in one region and this region would have the fastest rate of growth. There are various plausible mechanisms that would prevent all human capital engaged in research and development concentrating in a single region. One such would be if there were human capital generating institutions located in every region and only a small probability of any human capital element moving to another region within any given time period (an individual's working life). Another possible mechanism would be that research and development is sector specific and that there is some positive probability of a new sector emerging in any given region in each time period. Mechanisms such as these would, however, greatly complicate the Romer model and are beyond the scope of the present paper to explore. It might also be argued that the period analysed here is short enough (in the context of R&D facilities and personnel) to be consistent with a disequilibrium view anyway.

Other spatial formulations are, of course, possible but this suggests that if a regional measure of human capital employed in economically applicable knowledge generation can be found, then the prior expectation is that the influence that that would have on differential FUR growth rates would be raised to a power greater than 1. It should also be noted that it will tend to produce regional divergence since the agglomeration factor affecting the productivity of human capital employed in research will be a form of increasing returns to scale.

As an example of localised knowledge spillovers, when doing this research the author worked in a department which had long been interested in multinational firms and had collected a data base on all research and development establishments in Europe of Fortune top 500 firms, including their addresses. These were allocated to FURs and - to represent concentration of human capital employed in knowledge generation - expressed per unit population. While it is a crude measure - it would be more appropriate if the number of qualified researchers at each establishment were also known - it is a crude measure of the theoretically appropriate variable.

Including the log of the initial level of per capita income the full model, then, has the following form:

$$
\begin{aligned}
\dot{Y}_{Ft} = {} & \beta_0 + \beta_1 \ln Y_{Ft0} + \beta_2 \dot{Y}_N + \beta_3 C_{BG} + \beta_4 C_{NI} + \beta_5 \Delta PT + \beta_6 \Delta PT^2 \\
& + \beta_7 A_{75} + \beta_8 A_{75}^2 + \beta_9 Coal + \beta_{10} Port + \beta_{11} I_{75} \\
& + \beta_{12} \ln S + \beta_{13} L + \beta_{14} \sum \left(\frac{\dot{Y}_i - \dot{Y}_j}{D_{ij}} \right) + \beta_{15} \left(\frac{R}{S} \right)^n
\end{aligned} \tag{7}
$$

where:

$\dot{Y}_F$	=	percentage growth in FUR GDP per capita, converted at purchasing power parity to US dollars
$\ln Y_{Fto}$	=	log of mean FUR GDP p.c., 1979-82
$\dot{Y}_n$	=	percentage growth in GDP per capita in area of nation excluding major FURs, converted at purchasing power parity to US dollars
C_{BG}	=	dummy variable taking a value of 1 for all major FURs of Belgium and former Western Germany: zero elsewhere
C_{NI}	=	dummy variable taking a value of 1 for all major FURs of Northern Italy south to and including Rome
ΔPT	=	change in economic potential resulting from abolition of tariffs, EC/EU enlargement and reduced transport costs: calculated from Clark *et al* (1969) and Keeble *et al* (1988)

A_{75}	=	proportion of total labour force in agriculture in the wider Level 2 region in 1975
Coal	=	dummy variable measuring physical coincidence of FUR territory to coalfield: Coal = 2 if whole area of FUR coincident with area of coalfield; Coal = 1 if part of area of FUR coincides with area of coalfield
Port	=	dummy variable ranging from 0 to 4 scaled strictly to volume of seaborne trade through port in 1966, except for inland port of Duisburg, the largest on the Rhine
I_{75}	=	proportion of total labour force in industry in the wider Level 2 region in 1975
S	=	size of FUR measured as population in 1981
L	=	density of development within FUR, measured as population density in 1981
$\sum\left(\frac{\dot{Y}_i - \dot{Y}_j}{D_{ij}}\right)$	=	Sum of difference between growth rate of per capita income in FUR i and growth rate of per capita income in other FUR(s) j within 80 kms, divided by D, distance between FUR and other FUR(s) in kms. If a national boundary lies between FURs, actual distance is doubled
R	=	number of research and development establishments of 500 largest world industrial enterprises (Fortune listings) per million population; *n* to be determined by investigation

$\dot{Y}_F$ and $\dot{Y}_N$ were measured as percentage increase in income between the mean of 1979-82 and the mean of 1987-90. $\dot{Y}_i - \dot{Y}_j$ - the difference between the percentage change of income within a FUR and each of its neighbours - was measured between the mean of 1979-81 and the mean of 1983-86.

7 The Results

This model was estimated using OLS for the 118 FURs for which there were data. The results are shown in Table 1. Variables are added in groups, starting with the log of per capita income at the start of the period. The diagnostics of the full model were satisfactory with no indicated problems of functional form, non-normal residuals[16] or heteroscedasticity. The adjusted R^2 at 0.60 ($R^2 = 0.65$) for the

---

[16] Contiguity, linear, road and time distance measures of spatial interdependence were tested. Of these, that conforming to the most obvious measure - the inverse of time-distance - proved the most sensitive to changes in model specification. The full model

full model (Model 8) was satisfactory for a large cross sectional data set and all coefficients had the expected sign; most were significant at the 1% level and all were at the 10% level. A large number of experiments were conducted with alternative models and specifications and reestimation on sub-samples. With the notable exception of the initial per capita income level, the estimates for most variable parameters proved extremely robust and no sign ever changed. The general conclusion from these experiments was that the model performed well and that there were few underlying statistical or specification problems except for the convergence measure. In the more fully specified models presented in Table 1 (models 5 to 8) no coefficients changed by more than one standard error.

The results on the research and development variable are particularly noteworthy and were subject to considerable investigation. The performance of the variable when raised to a power of four was superior to any alternative although the t value never fell below 3 when the variable was raised to any power between 1 and 4. Nevertheless a power of 4 is a higher value than might have been expected. Investigation showed there was a very high leverage on one observation, which had the highest value on the R&D variable. To check that this was not conditioning the results, the model was reestimated excluding that observation. The form of the influence of the R&D variable was sensitive to this. The best performance was still obtained with the variable raised to a power greater than 1 but squared rather than to the power of 4. Omitting the outlying observation only trivially affected the overall performance of the model as a whole in terms of t-statistics for other coefficients, log-likelihood or adjusted R^2. The significance of the R&D variable, however, fell quite sharply. This implies that the particular result reported in Table 1 is significantly conditioned by the outlying observation but that the best fit, even excluding that information, is still obtained with the R&D variable squared.

---

with spatial interdependence specified *a priori*, that is with dummies for Northern Italian FURs and for FURs in Belgium and Germany (Model 8 in Table 1), performed best. The implication was that significant, but difficult to measure independent variables, affecting all FURs in those areas equally, were omitted. Using the Spacestat package the value for the Lagrange Multiplier test for a spatial lag (Burridge, 1980) in this model was 0.6151 which was not significant. An alternative model, omitting the spatial dummies and including a generalised spatial lag variable performed almost but not quite as well. If, however, either or both of the spatial dummies was omitted, there was clear evidence of spatial autocorrelation in the residuals of the estimated model.

Table 1 **Testing for Convergence/Divergence 1980-90: Dependent Variable % change in GDP p.c. @ PPS 1979/82 to 1987/90**

	Model 1	Model 2	Model 3	Model 4	Model 5	Model 6	Model 7	Model 8
1. Constant $\hat{\beta}$	120.2791**	49.0787	-28.7739	-18.8928	-49.0435	0.5783	29.9179	26.6780
t	(3.99)	(1.49)	(-0.73)	(-0.39)	(-0.84)	(0.01)	(0.62)	(0.59)
S.E.	30.1262	32.8683	39.6735	48.6610	58.4365	54.6070	48.2912	45.2406
2. log of GDP p.c. 1979-82	-6.6885*	-2.4741	4.9745	4.0857	5.9304	0.7541	-4.3033	-4.3931
	(-2.000)	(-0.76)	(1.25)	(0.76)	(1.00)	(0.14)	(-0.86)	(-0.93)
	3.3368	3.2788	3.9943	5.3675	5.9100	5.5623	5.0244	4.7073
National factors								
3. % Income growth 1980-90 in national remainders		0.4443**	0.5974**	0.5640**	0.7003**	0.7787**	0.7043**	0.6991**
		(4.21)	(5.74)	(4.98)	(4.80)	(5.75)	(5.84)	(6.19)
		0.1055	0.1041	0.1132	0.1459	0.1355	0.1206	0.1130
4. Dummy for Northern Italy			5.9234*	5.9764*	4.3462	4.2150	4.5171*	4.8716*
			(2.05)	(2.05)	(1.44)	(1.49)	(1.72)	(1.97)
			2.8918	2.9103	3.0239	2.9628	2.6340	2.4694
5. Dummy for Belgium and Germany			-5.8532**	-4.9362*	-4.9258*	-5.2045*	-4.5388*	-4.2842*
			(-2.88)	(-2.15)	(-2.12)	(-2.36)	(-2.26)	(-2.28)
			2.0351	2.2925	2.3211	2.2053	2.0068	1.8812
European integration								
6. Change in Economic Potential resulting from European integration				3.8852*	3.1866	8.7886	11.4841	12.4183*
				(0.51)	(0.40)	(1.11)	(1.64)	(1.89)
				7.5872	7.9996	7.9211	7.0100	6.5718
7. Change in Economic Potential resulting from European integration squared				-3.6083	-2.6684	-4.4654	-6.3947*	-6.5880*
				(-0.83)	(-0.61)	(-1.04)	(1.69)	(-1.85)
				4.3258	4.3681	4.2868	3.7915	3.5524
Objective 1 type variables								
8. % of L.F. in agriculture in 1975 in NUTS 2 region					0.7512*	0.4916*	0.4305	0.5523*
					(2.42)	(1.70)	(1.62)	(2.20)
					0.3106	0.2896	0.2657	0.2508
9. % of L.F. in agriculture in 1975 in NUTS 2 region squared					-0.0215*	-0.0188*	-0.0176*	-0.0197**
					(-2.57)	(-2.43)	(-2.52)	(-3.00)
					0.0084	0.0077	0.0070	0.0066
Objective 2 type variables								
10. Historic influence of coal industry						-5.1224**	-3.9444**	-3.3911**
						(-3.80)	(-3.28)	(-2.99)
						1.3483	1.2030	1.1358
11. Historic influence of port industry						-1.3513*	-1.4837**	-1.6508**
						(-2.37)	(-2.90)	(-3.44)
						0.5707	0.5108	0.4805
12. % of L.F. in industry in 1975 in NUTS 2 region						-0.1687	-0.1407	-0.1649*
						(-1.43)	(-1.35)	(-1.69)
						(0.1178)	(0.1040)	(0.0977)
Spatial processes								
13. Log of total population 1981							3.0428**	3.5551**
							(3.11)	(3.84)
							0.9778	0.9254
14. Growth differential with neighbouring FURS 1979-86 divided by distance							6.7084**	6.4082**
							(4.32)	(4.40)
							1.5519	1.4560
15. Population density 1981							-0.0017*	-0.0016*
							(-1.82)	(-1.78)
							0.0010	0.0009
Dynamic knowledge spillover effects								
16. Research and Devlpt Establishments per million, 1988 raised to power of 4								0.0038**
								(3.92)
								0.0010
Adjusted R^2	0.025	0.15	0.27	0.26	0.29	0.40	0.59	0.60

Note: for each independent variable the $\hat{\beta}$, t-statistic and Standard Errors are shown in order: ** significant at 1%; * significant at 10%.

8 Discussion and Conclusions

Apart from the measure of convergence, a quite full and robust model of differential income growth rates in the EU's major city regions is, therefore possible. It has interesting implications in policy terms, in terms of understanding the significant role and form of purely spatial factors and adjustment processes, and as providing evidence consistent with a spatial application of Romer's ideas (1990) in relation to dynamic increasing returns to human capital employed in knowledge production. In addition, the form of the spatial adjustment mechanism itself has significant welfare implications which policy should take into account.

Of the policy variables the result on the measure of the spatial effects of European integration is of most interest, not only in itself and the interpretation it has for the impact of European integration on 'core' compared to 'peripheral' regions, but in the context of previous results. Clark *et al* (1969) calculated regional economic potential on four assumptions; for the regions of the European Economic Community (EEC) of 6 before the Treaty of Rome; as it was for the EEC of 6 with tariff reductions following the Treaty of Rome; as it would be for the EEC of 10 after the first enlargement; and as it would be following that enlargement and a fall in transport costs associated with containerisation and roll-on roll-off ferries. Each value of regional economic potential used the same GDP data - those estimated by Clark and his associates for 1966. Thus the values were for different hypothetical states of a static world.

Seven studies using four unrelated dependent variables have now been undertaken by the present author. As the data base has been enlarged it has been possible to increase the range of dependent variables from simple percentage population change, to change in unemployment rate, to change in the value of the urban problem index for three periods between 1971 and 1987 and, now, for the percentage change in GDP per capita from 1979 to 1990. The results from these seven separate models with respect to the spatial impact of European integration are summarised in Table 2. In each case experiments were undertaken to get the best model performance.

The pattern is striking. Not only did a plausible relationship emerge for each model but for each period studied, irrespective of which growth variable (inverse growth in the case of change in unemployment rate) was the focus of analysis, the most appropriate form of the change in economic potential variable performed best. For the 1960s the estimated change associated only with tariff barrier reductions between the original six member states performed best. For later periods the estimated change associated with tariff reductions, transport cost reductions and enlargement performed best. Not only that, but the relationship moved from being linear, with the greatest gains being associated with the highest values right at the centre of economic gravity of the EU, to being quadratic, for the two most recent periods, suggesting deconcentration and spread effects.

Table 2 The role of changing Economic Potential: Estimated coefficients for different models

Measure of European Integration Effect	Periode over which Dependent Variable was Measured: 1961-71	1971-81	1977/81-1985/7	1971-78[a]	1975-83[a]	1975-87[a]	1979-1990
	Dependent Variable of Model: % Population Change[b]		Change in Unemployment[c]	Change in Urban Welfare Index[d,e]			%Change in GDP p.c.[f]
Change in Economic Potential following:							
i) Treaty of Rome	+ 0.0038*						
ii) All changes		+ 0.0038**	- 1.621*	+ 1.0780*	+ 1.3291*	+ 5.6145**	+ 10.2756
iii) All changes squared						- 2.6290**	- 5.5536

* Significant at 5%   ** Significant at 1%
[a] Calculated as a moving average so periods are not precise and overlap
[b] Reported in Cheshire & Hay 1989
[c] Reported in Cheshire 1991
[d] Reported in Cheshire 1993
[e] A linear form was reported for the whole period 1971-88 in Cheshire, 1990
[f] Reported in table 1.

Map 1 shows the changes in economic potential associated with tariff and transport cost reductions and enlargement. Identifying this as a measure of the spatial impact of European integration we can identify how the implied areas of maximum positive impact of European integration have moved outwards from the 'core' over time; that is as we move from the linear relationships estimated for models analysing data from 1961 to about 1983, to the quadratic forms estimated using data since 1975 (see Table 2 and footnote a). These regions of maximum favourable impact of European integration are shown in Maps 2a-c.

Map 2: Maximum favourable impact of change in economic potential variable on regional development

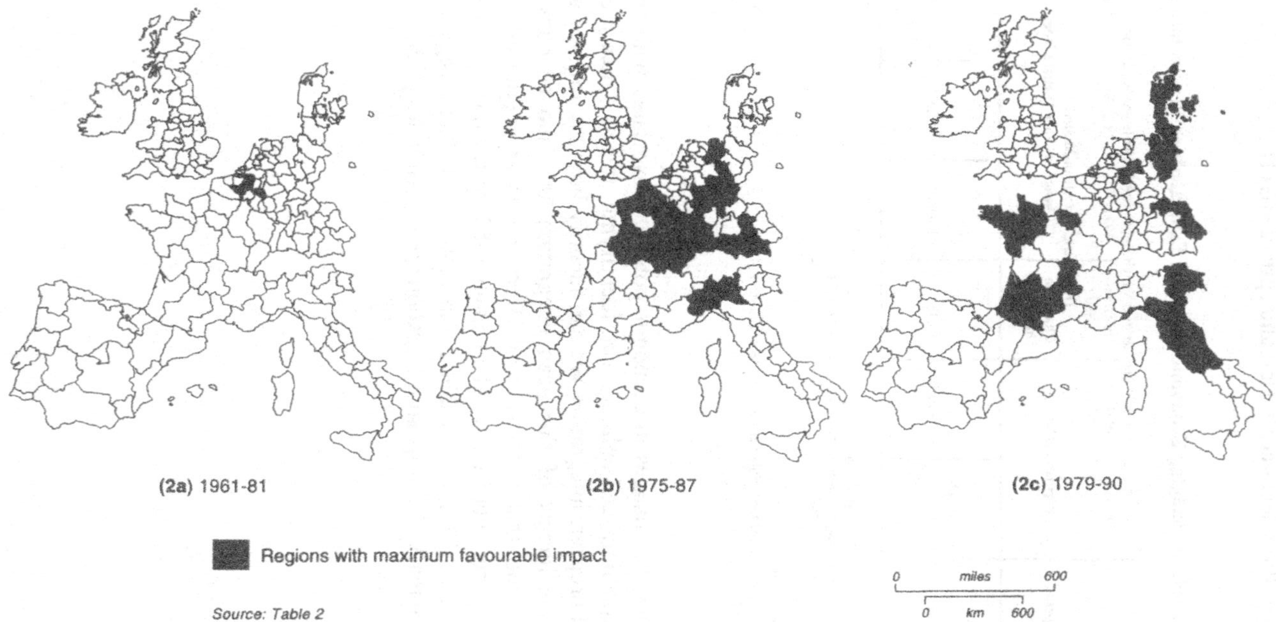

Source: Table 2

In interpreting these results it may be worth repeating two points. Clark *et al's* (1969) calculations of the impact on regional economic potential for different states of European integration and levels of transport costs used GDP data for a single date, 1966; in addition the gains in economic potential predicted tended to fall away with distance from the 'core' of the EU, that is from the Benelux countries and north western parts of Germany. These gains, however, tended to induce growth and with that growth, rising costs, not accounted for in Clark *et al's* static equilibrium framework. In addition there were continuing falls in transport costs unanticipated by Clark which extended the area of greatest gains from integration (see Vickerman, 1995, for a further discussion of how these more recent changes in transport costs might have been expected to influence spatial patterns of growth). Over time, therefore, the maximum favourable impact of the change in economic potential, as estimated by Clark, extended outwards from the core (to regions with gains in economic potential less than the highest estimated by Clark *et al* on the basis of 1966 data) and the impact of changes in economic potential as calculated by Clark *et al* became less important in determining regional growth differentials. For the latest period analysed (1979/82 to 1987/90) not only had changes in regional economic potential become only marginally significant but the greatest gains had spread out far beyond the traditional 'core' regions to areas previously considered 'peripheral' (see Maps 2a-c).

The coefficient on the spatial adjustment mechanism is also worthy of note. This variable contributes substantially to the performance of the full model and suggests that the observation of Mr. Millan, the Commissioner for Regional Policy, quoted above, does not relate to a chance factor but to a process which is wholly systematic and integral to the pattern of relatively dense urbanisation common in Europe (but not in the US)[17]. It also has interesting welfare implications which cannot be fully explored here. The implication has to be considered, however, that since quite low cost adjustment is possible via commuting pattern changes between closely clustered FURs with divergent growth rates, economic opportunities for characteristic specific groups in such cities tend to converge even while mean GDP per capita diverges. This would not be true for divergences in growth rates for more isolated city regions (except in as far as in the very long run migration,

[17] Given the arguments advanced above as to the processes of adjustment underlying this result, it also suggests more detailed and direct tests that could be performed given the availability of time series data on commuting, unemployment rates by skill level and employment by workplace, for areas small enough to be reconstituted into FURs. Such data exist for a number of European countries but would have to be collected from national sources and subject to significant further adjustment to give adequate comparability. Such a task would require resources approximately equal to those needed to collect all Census data to update FUR population data to 1990/91; that is resources on approximately the same scale again.

technology transfer, policy, capital movements or even diminishing returns to capital, might produce convergence).

That sanguine conclusion is, however, made less plausible by the results which are consistent with Romer's growth model (Romer, 1990). The results reported here, on balance, would seem to provide a more direct test of Romer's ideas about economic growth than do those of Glaeser *et al* (1992) who suggested that their evidence tended to support other theories. The dependent variable analysed in this paper is more appropriate than employment growth in city-industries on which Glaeser *et al* mainly relied and even though the measure of human capital employed in R&D is crude, it is at least a measure of the theoretically appropriate variable whereas Glaeser *et al's* analysis used local industrial concentration as a proxy. Some caution is appropriate given the evidence that the leverage of one outlying observation plays a non-trivial part in the specific result reported in Table 1. But as was noted above, if that observation is dropped, still the R&D variable works best when squared. Not only are these results consistent with the central mechanism of Romer's model, therefore, but they also support a spatial interpretation of it; that there are increasing returns to the concentration of human capital employed in R&D.

Alone this would suggest that per capita income growth rates might persistently diverge because of localised knowledge spillovers and dynamic increasing returns to knowledge. However, since the determination of net changes in GDP per capita is a multivariate process, this does not necessarily imply that observed per capita incomes would diverge.

In the present context, however, interest is on the result on the convergence variable itself. In the simple regression (Model 1) there is apparent strong and significant convergence in regional incomes over the 1980s. As soon as the growth rates in the national remainders are added, however, the convergence variable ceases to be significant. The most plausible explanation is that there is less variation in within-country growth rates than in between-country growth rates and the initial level of income is effectively acting as a national dummy in Model 1. As soon as the national growth variable and the other national dummies, representing *a priori* knowledge, are added, the convergence variable switches signs and we observe apparent but not significant divergence. Apparent divergence continues to be registered until Model 7, in which the spatial process variables are entered. In that, and in the full model, there is again convergence but the coefficient is non-significant.

These results seem amply to justify the scepticism expressed in the introduction. The most plausible interpretation would seem to be that there are some forces producing convergence and others producing divergence and the actual outcome over time is determined by the net effect of those forces. The estimate of the 'convergence' coefficient is highly sensitive to the specification of the model. It does not seem, therefore, that it is helpful to attempt to measure convergence in the way that has been done to date. It is perfectly reasonable to study the actual distribution of regional incomes, and changes in the pattern of the dispersion of regional incomes over time (Quah, 1995; Magrini, 1995). This is a valid dimension of equity with which policy, at least in the EU, is specifically concerned. But in analysing regional growth it is much more useful to try and develop theories and devise ways of rigorously testing such theories, than to provide measures of 'convergence' which are at best unreliable and perhaps entirely spurious.

References

Alonso, W. (1964), Location Theory in „Regional Development and Planning", (J. Friedmann and W. Alonso, eds.), MIT Press, Cambridge Mass.

Armstrong, H.W. (1994), Regional Growth within the European Union: a reappraisal of the cross-sectional evidence, paper to the 34th Congress of the European Regional Science Association, Groningen.

Armstrong, H.W. and J. Taylor (1993), Regional Economics and Policy, Harvester Wheatsheaf, London.

Barro, R. (1991), Economic growth in a cross section of countries, *Quarterly Journal of Economics*, 106, 407-501.

Barro, R. and X. Sala-i-Martin (1991), Convergence across states and regions, *Brookings Papers on Economic Activity I*, 107-82.

Barro, R. and X. Sala-i-Martin (1992), Convergence, *Journal of Political Economy*, 100, 223-51.

Baumol, W.J. (1986), Productivity growth, convergence and welfare: what the long run data show, *American Economic Review*, 76, 1072-85.

Blanchard, O. (1991), Convergence across states and regions: discussion *Brookings Papers on Economic Activity I*, 183-92.

von Böventer, E. (1975), Regional growth theory, *Regional Studies*, 12, 1-29.

Borts, G.H. and J.L. Stein, 1964 *Economic growth in a free market*, Columbia University Press, New York.

Bowers, J.K., P.C. Cheshire, A.E. Webb and R. Weeden (1972), Some aspects of unemployment and the labour market, 1966-1971, *National Institute Economic Review*, 54, 44-63.

Burda, M.C. (1993), The determinants of East-West German migration: some first results, *European Economic Review*, 37, 452-61.

Burridge, P. (1980), On the Cliff-Ord Test for Spatial Autocorrelation, *Journal of the Royal Statistical Society B*, 42, 107-8.

Cheshire, P.C. (1979), Inner areas as spatial labour markets: a critique of the inner area studies, *Urban Studies*, 16(2), 29-43.

Cheshire, P.C. (1990), Explaining the Recent Performance of the European Community's Major Urban Regions, *Urban Studies*, 27(3), 311-333.

Cheshire, P.C. (1991), Problems of regional transformation and deindustrialisation in the European Community, *Industrial Change and Regional Economic Transformation: the Experienc of Western Europe* (L. Rodwin and H. Sazenami Eds.), Harper Collins Academic, London.

Cheshire, P.C. (1993), Some Causes of Western European Patterns of Urban Change, 1971-88, *Urban Change in the United States and Western Europe: Comparative analysis and policy* (A. Summers, et al Eds.), The Urban Institute Press, Washington D.C.

Cheshire, P.C. (1995a), A New Phase of Urban Development in Western Europe? An analysis of 1990/91 Census data for EU countries, *Urban Studies*, 32(7), 1045-63.

Cheshire, P.C. (1995b), European Integration and Regional Response, *The regions and the Europe* (Methodes Ed.), Manchester University Press, Manchester

Cheshire, P.C. and D.G. Hay (1989), *Urban Problems in Western Europe: an economic analysis*, Unwin Hyman, London.

Clark, C., F. Wilson, and J. Bradley (1969), Industrial location and economic potential in Western Europe, *Regional Studies*, 3, 197-212.

Evans, A.W. and R. Richardson (1981), Urban Employment: interpretation and additional evidence, *Scottish Journal of Political Economy*, 282, 107-24.

Fothergill, S. and G. Gudgin (1982), *Unequal growth: urban and regional employment change in the United Kingdom*, Heinemann Educational, London.

Glaeser, E. L., H. D. Kallal, J. A. Scheinkman and A. Shleifer (1992), Growth in cities, *Journal of Political Economy*, 100, 1126-1152.

Gordon, I. and D. Lamont (1982), A model of labour-market interdependencies in the London region, *Environment and Planning A*, 14, 238-64.

Gordon, I.R. (1985), The cyclical sensitivity of regional employment and unemployment differentials, *Regional Studies*, 19, 95-109.

Gyourko J. and J. Tracy (1991), The structure of local public finance and the quality of life, *Journal of Political Economy*, 99, 774-806.

Hall, P. and D. Hay (1980), *Growth centres in the European urban system*, Heinemann Educational, London.

Hughes, A. (1993), Emerging settlement patterns and their implications for inner city poverty in the United States, *Urban Change in the United States and Western Europe: Comparative analysis and policy*, (A. Summers, et al., Eds.), The Urban Institute Press, Washington D.C.

Kaldor, N. (1970), The case for regional policies, *Scottish Journal of Political Economy*, 17 (3), 337-48.

Keeble, D., P. L. Owens and C. Thompson (1983), The urban-rural manufacturing shift in the European Community, *Urban Studies*, 20, 405-18.

Keeble, D., J. Offord and S. Walker (1988), *Peripheral Regions in a Community of Twelve Member States*, Office of Official Publications, Luxembourg.

Krugman, P. (1991), Increasing returns and economic geography, *Journal of Political Economy*, 99, 483-99.

Magrini, S. (1995), Economic Convergence in the European Union: A Markov Chain Approach, Discussion Papers in Urban and Regional Economics (Series C) No. 111, Department of Economics, University of Reading.

Marston, S. T. (1985), Two views of the geographic distribution of unemployment, *Quarterly Journal of Economics*, 100, 57-79.

McCombie, J. S. L. (1989a), A Synoptic View of Regional Growth and Unemployment: I. The Neo-classical theory, *Urban Studies* 25, 267-81.

McCombie, J. S. L. (1989b), A Synoptic View of Regional Growth and Unemployment: II. The Post-Keynesian theory, *Urban Studies* 25, 267-81.

Millan, B. (1993), Memorandum to the Informal Council of Ministers responsible for Regional Policy and Spatial Planning, Nov. 1993, mimeo.

Myrdal, G. (1957), Economic development and underdeveloped regions, Methuen, London.

Quah, D. (1993), Empirical cross-section dynamics in economic growth, *European Economic Review*, 37, 426-34.

Quah, D. (1995), Paper to Economic and Social Research Council Urban and Regional Economics Workshop, University of Reading, England.

Richardson, H. W. (1973), *Regional growth theory*, Macmillan, London.

Romer, P. M. (1986), Increasing returns and long-run growth, Journal of Political Economy, 94, 1002-37.

Romer, P. M. (1990), Endogenous Technological Change, *Journal of Political Economy*, 98, 71-101.

Solow, R. M. (1956), A contribution to the theory of economic growth, *Quarterly Journal of Economics*, 70, 65-94.

Sala-i-Martin, X. (1994), Cross-sectional regressions and the empirics of economic growth, *European Economic Review*, 38, 739-41.

Vickerman, R. W. (1995), The regional impacts of Trans-European Networks, *Annals of Regional Science*, 29, 237-54.

Wilson, W. J. (1987), *The truly disadvantaged: the inner city, the underclass, and public policy,* University of Chicago Press, Chicago, 1987.

The Regional Economic Structure of the European Union: an Analysis of Long-Term Developments

Willem Molle*
Netherlands Economic Institute (NEI)
Erasmus University
P.O. Box 4175
NL - 3006 AD Rotterdam

1 Introduction, Summary and Conclusions

In the present paper** we describe the changes that have occurred in the structure of employment by branch in the regional system of the EU over the period 1950 - 1990. We have distinguished four ten-year subperiods. Moreover, for some elements of the analysis we give a forecast for the year 2000. So we span half a century.

Our analysis of regional development in the European Union focuses on the core-periphery dichotomy. In the present study we have distinguished six types of region defined by their degree of centrality/peripherality. The more peripheral regions are, the more they appear to be characterised by low income-per-head levels, which in turn are the consequence of the relative concentration in these regions of low-value-added production sectors. Our main aim is to picture the way in which the industrial bases of the various parts of the system have evolved over the past 40 years.

The present analysis is split up into two parts.

– First we describe on the macro level the general development of the employment per branch; we find that the rate of structural change of branches at the EU level is about stable through time. Next we analyse for each branch the development of its spatial concentration and find that for most branches this indicator decreases continuously through time. Much in line herewith we find a very general

* The author wants to thank Martijn Blom and Eugene Nefkens for the support they have given to this paper by gathering the data and carrying out the calculations.

** The analysis we present in this paper is one in a series of studies of European regional-development problems performed by associates of NEI. The present analysis merely describes patterns with the help of fairly simple analytical devices. In the near future we plan to carry out a more in-depth analysis by using econometric tools.

trend of de-specialisation of almost all European regions. Indeed, the economic structures of all EU regions have become very similar.

- Second, we compare the growth performance of the different categories of region. We observe that wealth levels have continuously converged. We find moreover, that this has gone hand in hand with a continuous change in the regional concentration of low- and high-value-added branches of activity. On one hand the central regions have constantly performed less well than could be expected from their industrial structure. On the other hand peripheral regions have constantly improved their industrial structure. Intermediate regions tend to take intermediate positions on this scale.

The explicit objective of regional and cohesion policies in Europe is to limit the differences in wealth. The method to make poor regions catch up with the wealthy ones is to facilitate the development of higher-value-added productions in these regions. To that end various policies are pursued to improve the production environment of these regions. In the past, cohesion policies have become progressively stronger. Together with the autonomous development that has led the industrial structure of the various regions and hence their wealth levels to converge.

2 Simple Indicators

2.1 Employment Growth by Branch for the Whole of the EU

The changes in the overall industrial structure of the EU are well documented. The data (see table 1) can be analysed on two levels. On the level of broad sectors they show clearly the well known trends of the decline in agriculture, the rise and fall of manufacturing, and the continuous rise of the service sector. On the level of the individual branches we see some well-known developments too.

In the manufacturing sector, several patterns can be distinguished. At one end of the scale we find two branches in which employment decreased consistently over the whole period of analysis: Fuel and power products and Textile clothing and footwear. Three other branches started decreasing in the 1960s: Basic metals, Non-metallic minerals, and Food. Three branches start to decline in the 1970s: Chemical industries, Metal working, and Transport equipment. At the other end of the scale we find two branches that have continued to show a fair dynamism: Paper and Printing have stagnated in employment only since the 1970s: the mixed branch of Other Industry shows constant growth.

In the service sector the pattern is less diversified: all branches show constant growth. The difference is only in the growth rhythm. Most dynamic are Other market services, Financial services and the Non market services (threefold growth). Less dynamic are Commerce and Transport.

Table 1 Development of total employment by branch 1950-1990 (Millions)

Branch	NACE CODE	1950	1960	1970	1980	1990
Agriculture, forestry and fishery products	B01	32,4	24.7	15.4	12.3	8.5
Fuel and power products	B06	3,0	3.3	2.6	2.1	1.7
Ferrous and non-ferrous ores and metals,other than radioactive	B13	1,5	1.9	1.9	1.4	1.0
Non metallic minerals and mineral products	B15	1,9	2.3	2.2	1.9	1.6
Chemical products	B17	1,4	1.9	2.3	2.1	2.0
Metal products, machinery, equipment and electrical goods	B24	6,6	9.4	11.0	10.6	10.6
Transport equipment	B28	2.9	3.6	4.1	3.5	3.0
Food, beverages and tobacco	B36	3.6	4.2	4.1	3.7	3.5
Textile and clothing, leather and footwear	B42	8.3	7.8	6.6	5.0	3.6
Paper and printing products	B47	1.7	2.2	2.5	2.5	2.5
Products of various industries	B50	3.2	3.5	3.7	3.8	3.9
Building and construction	B53	7.4	9.7	11.2	10.3	10.2
Recovery, repair, trade, lodging and catering services	B58	14.7	18.5	20.9	24.9	26.8
Transport and communication services	B60	6.4	7.1	7.4	8.2	8.1
Services of credit and insurance institutions	B69	1.3	1.9	2.8	3.5	4.1
Other market services	B74	4.2	5.9	8.2	15.9	24.4
Non-market services	B86	8.4	10.5	15.0	20.3	24.2
TOTAL		109.1	118.4	121.9	131.9	139.8

The first question to ask is whether the structural changes observed here for the entire 1950-1990 period varied significantly among sub-periods. One hypothesis could be that the changes in the structure triggered by the economic crisis of the second half of the 1970s were much worse than the ones at the upswing of the long cycle in the beginning of the 1950s. Our data do not support that hypothesis; as a matter of fact the coefficient of similarity[1] between the EU industrial structures measured in the various benchmark years is in all periods about 0.9. The rate of change indicated by this coefficient is much higher than the those prevailing in the first half of the century (see among others Paretti and Bloch 1956). This suggests that all through the period of study the EU has been marked by a fast and continuous structural adaptation.

The change is likely to continue into the future. That is evident from a recent ERECO study (ERECO 1995a), which forecasts the growth of the branches of activity in the coming years and hence the evolution of the structure of the European economy in the year two thousand.

2.2 Regional Distribution of Branch Activity

The change in the spatial structure of branch employment in the EU can be described by location coefficients.[2] These L coefficients indicate the concentration of a branch by the degree to which the branch distribution diverges from the distribution of total employment across the EU regions (See table 2).

---

[1] Similarity index:

$$SI_t = 100 - \frac{\sum_{i=1}^{I} |\sigma_t - \sigma_{t-1}|}{2}$$

$\sigma_{i.t}$ = share of sector i in total European employment
i = economic sector

[2] Location coefficient:

$$LC_{i.} = \frac{\sum_{r=1}^{R} |s_{ir} - s_{.r}|}{2}$$

$s_{ir} = e_{ir} / e_{i.}$ = share of region r in total European employment of sector i
$s_{.r} = e_{.r} / e_{..}$ = share of region r in total European employment

Table 2 Location coefficients of the 17 branches of economic activity 1950-1990

Branch		1950	1960	1970	1980	1990
Agriculture, forestry and fishery products	B01	0,29	0,33	0,38	0,37	0,36
Fuel and power products	B06	0,44	0,40	0,28	0,26	0,19
Ferrous and non-ferrous ores and metals,other than radioactive	B13	0,38	0,37	0,36	0,36	0,35
Non metallic minerals and mineral products	B15	0,17	0,15	0,15	0,17	0,20
Chemical products	B17	0,26	0,24	0,23	0,23	0,21
Metal products, machinery, equipment and electrical goods	B24	0,24	0,23	0,20	0,19	0,21
Transport equipment	B28	0,32	0,26	0,24	0,24	0,24
Food, beverages and tobacco	B36	0,13	0,12	0,11	0,14	0,14
Textile and clothing, leather and footwear	B42	0,20	0,19	0,21	0,24	0,33
Paper and printing products	B47	0,25	0,22	0,19	0,18	0,17
Products of various industries	B50	0,14	0,12	0,13	0,14	0,15
Building and construction	B53	0,10	0,08	0,08	0,08	0,07
Recovery, repair, trade, lodging and catering services	B58	0,14	0,09	0,06	0,06	0,06
Transport and communication services	B60	0,16	0,13	0,10	0,09	0,08
Services of credit and insurance institutions	B69	0,23	0,19	0,17	0,15	0,15
Other market services	B74	0,15	0,12	0,11	0,11	0,13
Non-market services	B86	0,12	0,11	0,10	0,10	0,09

The table clearly shows a trend towards less concentration for the majority of branches. The opposite occurs for the branches of Agriculture, Textiles, while for the branches Mining, Food and Other products no clear change pattern can be discerned.

2.3 Branch Structure of Regional Economies

When there is a trend towards lower concentration of branches, the degree to which regions are specialised in certain sectors is likely to show a decreasing trend too. For a more quantitative insight, we have calculated the specialisation coefficient for each region and benchmark year.[3] The results show that the following two types of development characterise the quasi-totality of the regions:

- continuous decrease; this is the case of the overwhelming majority of regions. To this group may be added the regions for which the coefficient rose in the 1950s and decreased consistently afterwards;
- no change; this was notably the case of regions had a low specialisation coefficient to begin with.

The same analysis has been carried out on the regions grouped in six classes of centrality (1 = most central, 6 = most peripheral (see Annex 2). The results are given in the next table.

Table 3 Specialisation coefficients (17 branches) of the economies of regional groups 1950-1990.

Group	Number	1950	1960	1970	1980	1990
1	6	0.23	0.19	0.14	0.12	0.09
2	14	0.11	0.10	0.08	0.07	0.06
3	15	0.07	0.07	0.09	0.08	0.06
4	20	0.08	0.06	0.06	0.03	0.03
5	23	0.07	0.07	0.07	0.06	0.05
6	18	0.31	0.30	0.24	0.20	0.18

The highest initial specialisations can be observed at either end of the scale ; for the central regions the decrease of specialisation was even more marked than for the peripheral regions.

The conclusion from this analysis on the levels of the individual region as well as the aggregate group is twofold:

- there is a very general trend towards de-specialisation of regions;
- the developments are largely continuous through time.

---

[3] Specialisation coefficient:

$$SC_{.r} = \frac{\sum_{i=1}^{I} |\sigma_{ir} - \sigma_{i.}|}{2}$$

$s_{ir} = e_{ir} / e_{r.}$ = Share of sector i in total employment of region r

$s_{i.} = e_{i.} / e_{..}$ = Share of sector i in total European employment

For easy understanding of the detailed results by individual region, we have abstracted from a number of details. We have grouped regions by categories of development, and compared only the situations of 1950 and 1990 (see Table 4).

This table gives rise to the following remarks grouped by level of specialisation in 1990:

- the low-specialisation regions, whatever there initial situation, are practically all in the heartland of North-Western Europe; among them are many regions for which the very high initial specialisation in manufacturing has dramatically decreased;
- the regions which still had relatively high specialisation levels in 1990 are all situated in the periphery of the EU; they are often specialised in the more traditional industries, and sometimes in new industries that have a relatively high weight in the (small) economy.
- the medium-specialised regions form a mixed bag. We find on the one hand the capital-city regions of the countries of North-Western Europe, and on the other some of the central regions of other countries; their adherence to this group is based on their relative specialisation in services. Next we find many of the intermediate regions whether in the heartland of the EU or in more outlying countries. Finally, some more peripheral regions also fall into this group; their medium specialisation is explained by the modernisation of their economy.

3 Composite Indicators

3.1 Convergence or Divergence?

Much has been written about the centre-periphery dichotomy in spatial economics. The centre is defined as a concentration of diverse economic activities and the nodal point of many communication links. The periphery groups the zones that are in general rather remote from the central zone, with a limited economic base and infrastructure. Of course, this dichotomy is not absolute: there are many intermediate regions that cannot be easily fitted into this mould (see Annex 2).

Much has also been written about the way in which the system, once a core-periphery contrast installed, is likely to develop. We can distinguish here two main strands of thought. The convergence school predicts that the movement of production factors will eventually even out the differences between the two types of region. The divergence school predicts that a system characterised by initial disparities and inadequately functioning markets will end up with greater differences (see for a short characterisation and some references to relevant literature, Molle 1995).

Table 4 Changes in the economic structure of the regions of the EU15 by category of specialisation coefficient in 1950-and 1990.

'50\'90	1: <10	2: 10-20	3: 20-30
1: <10	Schleswig-Holstein Niedersachsen Hessen Bayern Picardie Haute-Normandie Rhone-Alpes	Champagne-Ardenne Danmark Osterreich	
2: 10-20	Rheinland-Pfalz Centre Bourgogne Alsace Friuli-Venezia Giulia Vlaams Gewest East Anglia	Nordrhein-Westfalen Baden-Wurttemberg Franche-Comte Languedoc-Roussillon Provence-Alpes-Cote d'azur Piemonte Liguria Lombardia Trentino-Alto Adige Veneto Umbria Lazio Campania Noord Oost Zuid Ireland Pais Vasco Cataluna Sverige	Luxembourg Baleares
3: 20-30	Nord-Pas-de-Calais Pays de la Loire Aquitaine Yorkshire & Humberside East Midlands South West North West Scotland	Saarland Basse-Normandie Lorraine Bretagne Poitou-Charentes Midi-Pyrenees Limousin Auvergne Corse Emilia-Romagna Sicilia Sardegna West Region Wallonne North Ireland Cantabria Navarra Madrid Suomi	Bruxelles-Brussel Asturias Rioja Comunidad Valenciana Canarias Portugal
4: >30	North Wales	Hamburg Bremen Berlin (West) Ile de France Toscana Marche Abruzzi Molise Puglia South East West Midlands Aragon Castilla-Leon Murcia	Valle d'Aosta Basilicata Calabria Hellas Galicia Castilla-La Mancha Extremadura Andalucia

Which one of these two views is most realistic in the case of the EU? The suggestion from much previous work (for instance, CEC 1995; Molle 1995; Vickerman and Armstrong 1995) is that convergence is dominant in Western Europe. A continuous process of long-term convergence is also evident from the next table.[4]

Table 5 Evolution of the divergence of the regional GDP/P levels by peripherality class of region (index to EU average).

GROUP	1950	1960	1970	1980	1990
1	153	143	136	127	126
2	120	124	127	117	116
3	103	111	105	110	111
4	97	98	95	102	102
5	97	89	90	93	95
6	32	32	41	47	46

The convergence of the GDP/P levels of the various categories of peripherality is indeed marked for the two extreme categories, still clearly visible for the categories 2 and 5 and rather blurred for the intermediate groups (3 and 4).

3.2 Regional Patterns

We assume that the convergence of wealth reflects the restructuring of the spatial economy of the EU, by which the peripheral regions have acquired ever higher shares in high-value-added activities. To test that hypothesis we have carried out two sets of shift-share analysis, one for each of the 100-odd regions and one for each of the six peripherality categories of regions.

Some basic features can be observed from the data on the evolution of the structure of total employment presented in the left-hand part of table 6:

- *initial structure (share).* For all the subperiods distinguished, the central regions invariably show the highest shares, while the most peripheral regions invariably show the lowest shares. This means that the former have continuously transformed their economic base so as to be always in the best starting position

[4] Note that the process has levelled off since the 1980s; differences between the years 1980 and 1990 are statistically not significant. The explanation is still due. The cause may be the increased competition all regions and notably peripheral ones have had from extra EU producers.

for the next round. The opposite is true of the peripheral regions; notwithstanding continuous adaptation to new situations, they have always been least equipped for the next round. Intermediate regions took intermediate positions between the two extremes just depicted. Note also that the share component tends to decrease continuously with the degree of peripherality;

- *change (shift)*. The peripheral regions did much better over the whole period than could be expected from their structure; in all periods the figure for their shift is positive and larger than that of all other categories. The opposite seems to be true of the central regions, where the shift is mostly negative, indicating that part of their growth potential could not be exploited. A clear dichotomy can be observed in the last three periods of the analysis between the regions with an above-average centrality and those with an above-average peripherality, the latter showing a positive shift and the former a negative shift.

- *Structure vs change (share vs shift)*. The figures found for either component of regional growth show a distinct opposite pattern. (correlation coefficient, all periods pooled: 0.6).

The analysis of aggregates may hide a wide variety of individual patterns. To check whether that is so we have executed a shift-share analysis for all individual regions. For easy comparison of so many individual trajectories we have defined categories by the possible combinations shift and of share components; for instance, positive share/positive shift; positive share/negative shift, etc. The analysis has shown that the results for any category were highly representative of the results of the regions belonging to that category: indeed, a very large majority of the regions in each category revealed the same characteristics as the others in their group.

The 40-year trend is not likely to halt. Indeed, detailed calculations with the help of an econometric model show that the shifts in the location of major branches of activity is to continue up to 2000, and that the trend towards convergence of regional wealth creation may be expected to continue (ERECO 1995b).

Table 6 Total, share and shift effect of employment growth by branch and per category of region, 1950-1990, total and manufacturing.

1950-1960

	Total employment			Manufacturing		
Group	Total	Share	Shift	Total	Share	Shift
1	1,98	2,39	-0,41	0,92	1,18	-0,27
2	1,92	1,58	0,33	1,27	1,06	0,21
3	1,08	1,25	-0,17	0,93	0,93	-0,00
4	0,29	0,65	-0,36	1,03	1,00	0,03
5	0,45	0,63	-0,18	0,82	0,93	-0,12
6	0,34	-0,75	1,09	1,17	0,71	0,47

1960-1970

	Total employment			Manufacturing		
Group	Total	Share	Shift	Total	Share	Shift
1	3,95	5,19	-1,24	-0,04	1,82	-1,85
2	1,36	2,96	-1,60	0,12	1,45	-1,33
3	-0,49	1,65	-2,15	1,11	0,81	0,29
4	0,97	0,32	0,66	1,62	0,90	0,72
5	2,13	-0,23	2,37	2,20	0,49	1,71
6	-2,65	-5,55	2,90	2,13	-0,53	2,66

1970-1980

	Total employment			Manufacturing		
Group	Total	Share	Shift	Total	Share	Shift
1	0,30	1,66	-1,36	2,20	0,83	1,37
2	0,43	1,11	-0,67	1,39	0,94	0,45
3	0,74	0,88	-0,14	1,03	1,04	-0,01
4	2,30	0,99	1,31	-0,37	0,99	-1,37
5	1,11	0,95	0,17	0,62	1,10	-0,47
6	1,48	0,10	1,38	-0,11	1,27	-1,37

1980-1990

	Total employment			Manufacturing		
Group	Total	Share	Shift	Total	Share	Shift
1	1,37	1,91	-0,54	1,54	0,91	0,63
2	1,32	1,21	0,11	0,92	0,94	-0,01
3	0,78	0,82	-0,03	1,32	1,08	0,24
4	1,10	1,05	0,05	0,40	1,03	-0,63
5	0,98	0,96	0,02	0,75	0,96	-0,21
6	0,32	-0,30	0,62	0,59	1,10	-0,51

3.3 Branch Patterns

The shift-and-share analysis reported on above gives only aggregate results. It would be interesting to know whether the picture changes if we concentrated on specific sectors and branches. To that end we will carry out two types of analysis:

First we will look for distinct patterns in branch relocation. One hypothesis that could be tested is that of filtering down.[5] This is the spatial version of the well known life-cycle theory of products. At the early stages the development of a product (branch) will need much contact with clients and R and D activities and for that reason the entrepreneurs favour central locations for their production sites. At later stages the cost of production will become the dominant consideration, and the branch will be relocated to areas with lower labour cost. That development may at first favour intermediate regions, but later also peripheral regions.

The method adopted to test this hypothesis is a very simple one: we just compare the shares of each type of region in the total European employment of the branch in question through time.[6] Annex 3 gives the results.

The first analysis[7] of the the Annex shows two aspects:

- loss of the share of central regions and a gain of one or more of the other categories of regions.
- no clear-cut filtering-down processes in the sense that branches first move from category 1 to 2 and then on to 3 and 4 and finally to 5 and 6. This must not be understood as a rejection of the filtering-down process however, as an observed decrease of category 1 coupled with an increase of category 6 may actually be the net result of simultaneous moves from 1 to 2, from 2 to 3, from 3 to 4 from 4 to 5 and finally from 5 to 6.

A more in-depth analysis of the table shows that the branches may be grouped by five types of relocation pattern[8 9]:

---

[5] Both the division into branches and that into regions may be too aggregated for a successful test of this hypothesis. Moreover, the European space has become relevant for the relocation of manufacturing branches only in the 1960s and for service branches only in the 1980s (Molle 1994).

[6] This simple method is justified, as the shares in the total employment of each category of region are very stable through time. The only change of some consequence that should be remarked is the increase in the share of the central regions and the corresponding decrease in the share of the peripheral regions in the period 1950 - 1970. This period was indeed one of much immigration into the central and outmigration from the peripheral regions.

[7] In this analysis we disregard Agriculture (concentration in peripheral regions), Other market services (no significant changes) and Non-market services (government induced).

- From inner central (1) to outer central (2) regions. This is the case of Chemicals and for Paper. In later years more intermediate regions, for instance from category 4, also benefit from relocation of Chemical activity; while in the same period category-5 regions also benefit from an increase in Paper production.
- From central (1) to outer intermediate (4) regions. The exchange seems to pass the inner intermediate regions by. This applies to Machinery (later also moving to 5), Food (up to 1980 also moving to 5, later even to 6), Other manufacturing, and to Financial services (after 1980 also moving to 5).
- From central (1) to outer peripheral (6). This applies to two branches, viz Construction and Trade.
- From central (2) to inner and outer periphery (5 and 6). This pattern prevails for the branches Fuel and Ores and Basic metal.
- From all above-average central regions (1,2,3) to all above- average peripheral regions (4,5,6). This applies to Non-metallic minerals and Textiles.

Next, let us find out whether the convergent tendency observed for the structure of the whole economy apply also to the structure of the manufacturing sector. The analysis of the location coefficients in a previous section suggests that this is not the case. To look somewhat deeper into the matter, we have carried out a shift-and-share analysis on branches of the manufacturing sector. The results appear in the right-hand part of table 6.

We observe a reversal of trends in the course of our period of analysis. In the 1950s and 1960s, the convergence of structures was accelerating, with the regions that were worst placed for development of new branches (low share) growing fastest (high shift). In the 1970s and 1980s the industrial transformation that followed the economic crisis was carried through with much more positive effects for the central regions (high shift) than for the peripheral ones. This led to divergent developments. In the process, a substantial part of the earlier convergence was actually lost, with the result that over the whole period the structure of manufacturing has not converged.

---

[8] The branch of Transport equipment does not seem to fit this mould very well, as a matter of fact it shows a decrease in 1 and 4 and an increase in 3 and 6. May be the heavy government intervention in this branch for purposes of regional policy has given rise to this relatively complicated pattern.

[9] The results of the present analysis are complemented by similar ones performed for smaller areas on a lower regional aggregation. In that way Molle and Vianen (1981) have shown similar patterns for seven types of region in the Netherlands through four five-year periods (1960-1975). A pattern of convergence of regional branch structures was also found in a more recent study for the Netherlands (NEI/TNO/INRO 1995). The same study foresaw continuation of this trend for the coming decades.

4 Ways of Improvement

We have set ourselves to describe analytically the patterns of restructuring followed by the regional economies in the European Union over a long period of time. Simple devices applied to our unique data base have permitted us to identify some basic patterns in this restructuring. The analysis can of course be much improved in many respects by taking accounr of the factors that explain the patterns observed. We mention as such :

- Different branch patterns; to explain them would imply a more detailed analysis of the factors that determine location.[10]
- Different regional patterns; some regions have been more successful than others in restructuring their economy. This may in part be explained by preferences of industries for certain types of locational environment. For another part it may have sprung from the different capacity of regions to improve their competitive situation (see among others Brülhart 1995) possibly by upgrading their locational profile (see for an explanation of the relevant factors and of the role of regional policy, notably in education and infrastructure, De la Fuente and Vives 1995).

[10] A very detailed approach, using a intersectoral interregional model, was followed by Molle 1983. This Fleur-model explained for each branch of economic activity the differential growth of the employment in each region of the EC9 with the help of differential endowments of these regions with locational factors, such as market access, labour cost, urbanisation etc. However, the relative success of this approach in the explanatory stage was insufficient to build up a projection model. Given the high data demands of the model, the efforts to keep the model operational have been discontinued.

Annex 1 The data base

The systematic analysis of long-term developments of branches in the regions of the EU is hindered by the paucity of published data. To overcome that problem we have set up our own data base.

The major characteristics are :

Periods The data cover the period from 1950 to 1990. Basic data have been adapted to fit the standard benchmark years 1950, 1960, 1970, 1980 and 1990.

Areal coverage The data cover the present EU15. The EU started out as a relatively small grouping of countries in western Europe and has been gradually extended since. We have opted for the analysis of the present EU for the whole study period.

Regional division The standard nomenclature for regional divisions in the EU, called NUTS, distinguishes several levels of detail. For an analysis of centre-periphery problems, the second level would be appropriate. Data have been established on this level for France, Spain and Italy. In the other large countries, viz. the UK and Germany, this was not possible and we have worked with regions that are somewhat larger (such as the Laender in the FRG). Some countries could not be regionalised at all; these countries have been taken up as one region (Sweden, Finland, Denmark, Ireland, Portugal, Greece). We have not included data on the former East Germany, because evidently the different economic order that prevailed during the entire period of analysis precludes that (see map A1 for an overview of the division of the EU15-area into 96 regions).

Employment The only indicator for which data on economic activity by branch are available for a long period, is employment. The data for the present study come from three sources. The data on employment by branch and region for the nine countries of the EC for the years 1950, 1960 and 1970 have been derived from the framework of the Fleur study (see Molle 1983). The data on the other countries now included in the EU 15 for these early benchmark years are national data; only for Spain could a regionalisation be made. Finally, figures for the years 1980 and 1990 for all the regions and countries of the EU15 we have distinguished have been taken from various sources, including data published by Eurostat and many national statistics, some of which have been used also for the ERECO study on European Regional Prospects. Special care has been given to the comparability of definitions, both through time and across countries.

Branches

The NACE /CLIO of the EU gives guidelines for the division into branches of economic activity. This division is hierarchical. We have tried to establish data for the branch level (two-digit code), but have had to compromise, as on the level of the regional division chosen we could not construct data for all these categories that were comparable over space and time. The result finally obtained is based on the NACE branches and distinguishes some 17 categories, given in table 1.

GDP/P

The second and third indicators for which complete data have been gathered are GDP (total) and population. In this study we have used only the GDP/P indicator to depict the development of the relative wealth level of sets of regions distinguished.

Map A1 The regional division of the EU15

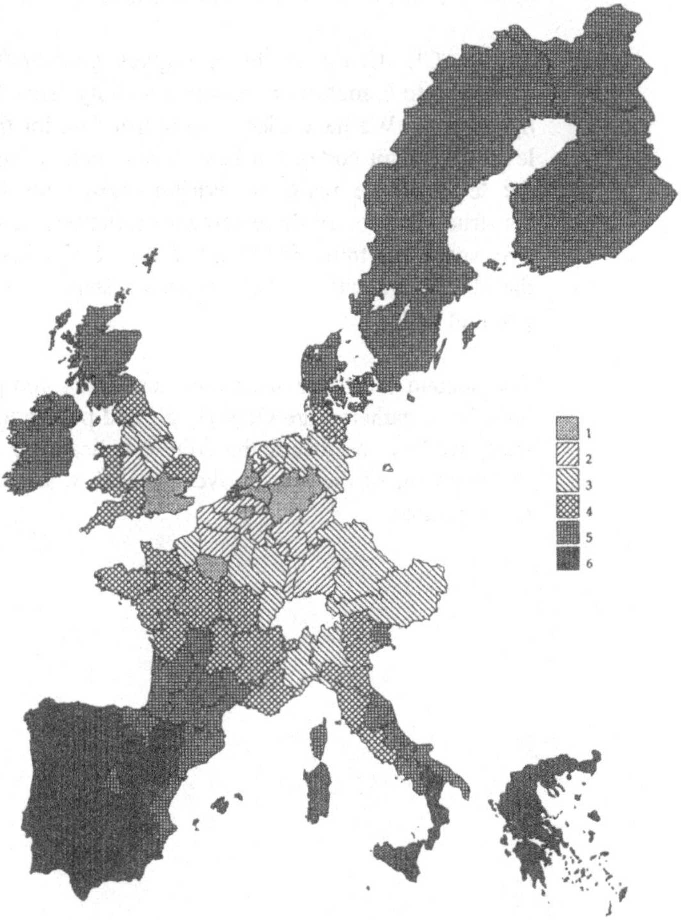

Annex 2 The definition of peripherality

The pattern of centrality and peripherality in Europe has been the object of many studies. For our work we had recourse to the study carried out by Keeble et al. (1988) for the European Commission. They defined an index of peripherality that measures access of the volume of economic activity of all other regions from the region in question. The volume of economic activity is measured by Gross Domestic Product. Accessibility is measured on the basis of two elements. First the road distance between all pairs of regions, account being taken of the extra distance cost involved in sea crossings to link island regions with continental ones. The second is the distance decay coefficient.

They have calculated the index for the year 1983 for the EU of 12 using a detailed regional division of the EU-12. The results of the calculations show very wide differences in peripherality in Europe. In the study these differences have been visualised by mapping regional groupings of regions with a similar degree of peripherality. One such grouping distinguishes three classes on either side of the EU-average (from class 1 most central to class 6 most peripheral). We have adopted that grouping, which divides the entire EU into groups of about equal size.

PM:
An alternative grouping was made in the Keeble et al. study, defining a basic threefold framework of central, intermediate, and peripheral regions; splitting up the first and last categories into inner and outer produces a categorisation into five classes (again from 1 inner central to 5 outer peripheral). The authors prefer the latter classification notably because it seems to take better account of the urban-rural dichotomy, so relevant to policy purposes. However, that advantage is lost in our regional division, which distinguishes only fairly large regions in most countries.

For the purpose of the present study we have not been able to make similar calculations based on our regional classification for each of the benchmark years and for the whole of the EU-15. This implies three things. First, we had to work with the 1983 classifications for the whole study period. Second, we had to attribute a peripherality-index classification to our regions on the basis of the dominant classification of the various constituting detailed regions distinguished in the Keeble et al. study. Third, we had to attribute to each of the three new EU members a classification based on expert judgement rather than calculation. The result of our set-up is given on Map A1.

Annex 3 Evolution of the shares of each category of region in total employment of the branch, 1950-1990 (%)

GROUP		1	2	3	4	5	6	Totoal
1950	B01	4	9	16	19	24	28	100
1960	B01	4	8	15	18	24	32	100
1970	B01	4	7	12	18	24	34	100
1980	B01	4	7	12	18	24	36	100
1990	B01	6	8	11	17	22	37	100
1950	B06	26	20	22	17	12	4	100
1960	B06	27	18	24	15	12	4	100
1970	B06	25	16	23	15	15	6	100
1980	B06	21	15	24	17	16	7	100
1990	B06	24	13	20	15	17	11	100
1950	B13	19	20	27	13	14	6	100
1960	B13	22	20	28	12	13	5	100
1970	B13	21	19	27	11	16	6	100
1980	B13	20	18	27	12	16	7	100
1990	B13	22	17	27	14	13	8	100
1950	B15	13	21	26	13	18	8	100
1960	B15	13	19	25	14	18	10	100
1970	B15	14	18	22	15	20	11	100
1980	B15	11	18	21	17	20	13	100
1990	B15	10	16	21	17	22	15	100
1950	B17	24	16	26	14	14	5	100
1960	B17	26	18	26	13	12	5	100
1970	B17	25	18	26	13	12	6	100
1980	B17	22	17	26	15	13	7	100
1990	B17	21	19	23	14	17	6	100
1950	B24	26	20	25	11	14	4	100
1960	B24	25	23	25	11	12	4	100
1970	B24	24	22	25	12	12	4	100
1980	B24	21	22	25	13	14	5	100
1990	B24	19	22	24	15	16	4	100
1950	B28	26	19	22	16	15	2	100
1960	B28	24	20	24	14	14	3	100
1970	B28	22	20	31	12	12	3	100
1980	B28	18	19	33	13	13	4	100
1990	B28	15	19	33	12	14	6	100
1950	B36	18	18	22	13	19	9	100
1960	B36	16	17	22	13	21	11	100
1970	B36	17	16	22	15	22	9	100
1980	B36	14	16	21	17	23	9	100
1990	B36	13	16	22	17	19	14	100
1950	B42	14	16	31	11	18	10	100
1960	B42	12	16	29	12	19	11	100
1970	B42	11	15	28	13	20	13	100
1980	B42	9	13	28	15	20	15	100
1990	B42	7	11	26	19	21	17	100

GROUP		1	2	3	4	5	6	TOTAL
1950	B47	30	14	21	11	20	4	100
1960	B47	28	16	22	10	19	5	100
1970	B47	28	16	21	11	19	5	100
1980	B47	24	15	22	12	21	6	100
1990	B47	23	18	21	14	19	5	100
1950	B50	18	16	22	11	20	12	100
1960	B50	17	16	23	13	20	11	100
1970	B50	16	15	22	14	21	12	100
1980	B50	14	15	22	16	22	12	100
1990	B50	13	16	23	18	20	10	100
1950	B53	19	16	21	16	20	9	100
1960	B53	18	15	20	16	20	11	100
1970	B53	18	14	18	16	22	12	100
1980	B53	16	14	19	17	20	15	100
1990	B53	16	13	19	16	20	16	100
1950	B58	23	15	21	15	18	7	100
1960	B58	22	16	22	15	17	7	100
1970	B58	21	16	20	16	19	9	100
1980	B58	19	14	19	18	19	10	100
1990	B58	19	14	20	17	19	12	100
1950	B60	22	15	21	14	19	8	100
1960	B60	23	16	21	14	19	9	100
1970	B60	22	15	19	14	20	11	100
1980	B60	21	15	18	16	19	11	100
1990	B60	22	15	18	15	20	11	100
1950	B69	33	13	18	12	16	7	100
1960	B69	32	15	19	12	15	6	100
1970	B69	33	16	19	12	15	6	100
1980	B69	30	14	18	15	15	8	100
1990	B69	28	15	19	14	17	7	100
1950	B74	26	14	20	15	18	8	100
1960	B74	26	16	19	14	17	8	100
1970	B74	26	16	17	14	19	8	100
1980	B74	25	15	17	17	17	9	100
1990	B74	24	15	19	17	18	7	100
1950	B86	22	14	19	16	18	9	100
1960	B86	22	15	19	16	19	9	100
1970	B86	22	15	19	15	20	9	100
1980	B86	20	14	18	16	22	10	100
1990	B86	19	14	18	17	22	10	100
1950	TOTAL	16	14	21	16	19	14	100
1960	TOTAL	18	15	21	15	19	13	100
1970	TOTAL	19	15	20	15	19	12	100
1980	TOTAL	18	15	20	16	19	12	100
1990	TOTAL	18	15	20	16	19	12	100

References

Brülhart, M. (1995), Trade and geographical specialisation in the European Union, paper for the AFSE Conference, Nantes

CEC 1995: Fifth periodic report on the Socio-economic situation and development of the regions of the Community, OOPEC Brussels

ERECO (1995a), Europe in 2000; economic analysis and forecast, ERECO, Brussels

ERECO (1995b), Regional Economic Prospects, ERECO, Brussels

Fuente, A. de la and X. Vives (1995), Infrastructure and education as instruments of regional policy; evidence from Spain, Economic Policy; an European Forum, Vol 20, pp 11 - 52

Keeble, D.E., 7 Offord and S. Walker (1988), Peripheral regions in a Community of twelve member states, CEC Luxembourg

Molle, W. and J.G. Vianen (1981), Werkgelegenheid, spreiding en verstedelijking, in L.H. Klaassen, W. Molle and J. Paelinck (eds) De dynamiek van de stedelijke ontwikkeling in Nederland, NEI Olijfreeks 1981/1 Rotterdam, pp. 130-155

Molle, W. (1983), Industrial Location and Regional Development in the European Community, the Fleur model, Gower Press, Aldershot

Molle, W. (1994), The Economics of European Integration; Theory, Practice, Policy, Dartmouth, Aldershot. (2nd ed)

Molle, W. (with I. Boeckhout) (1995), Economic Disparity under conditions of integration, a long term view of the European case, Papers in Regional Science, ; The Journal of the Regional Science Association International, Vol 74 pp 105-123.

NEI/TNO/INRO (1995), Ruimtelijke economische ontwikkeling, factoren in Nederland; Toekomstperspectief, MEZ, Den Haag.

Paretti, V. and G. Bloch (1956), Industrial production in Western Europe and the United States, 1901 to 1955, Banca Nazionale del Lavoro Quarterly Review, Vol 39 pp.

Vickerman, R.W. and H.W. Armstrong (1995), (ed) Convergence and Divergence among European regions; Pion, London

The Role of the Regions in the European Union - The Future of the Committee of the Regions

Arild Saether, Nicola Schmidt-Nissen, Kerstin Lorenz
European Institute of Public Administration
O.L. Vrouweplein 22
NL - BE Maastricht

1 Abstract

It is the purpose of this discussion paper to make a small contribution to the ongoing debate on the future of the regional level in the decision making process of the European Union.

The paper traces the evolution of the regional policy in what has today become the European Union. This brings us from the Treaty of Paris where no need for regional policies was recognized, through the establishment of the European Investment Bank in the Treaty of Rome, the development of regional tensions because of the Common Market, to the explicit recognition of the need for regional policies as part of the accomplishment of the Internal Market.

Thereafter it outlines the regional policy of what today is the European Union, the beginning, its objectives and instruments, the dramatic increases in its budgets, and discusses how the development of the regional policy of the Community has contributed to the creation of an identity of the regions.

The difference between a regional policy and a policy of the regions is then pointed out. The first is a development policy for the disadvantaged regions of the Member States.The second is a policy vis-a-vis the regions in which it tries to integrate the regional level of the Member States into the decision making process of the European Union.

Next the paper discusses the establishment of the Committee of the Regions, its role as an advisory committee, the composition of the Committee and its constituent bodies. In this section are also mentioned some of the opinions issued by the Committee.

The paper thereafter elaborates on other forms of regional participation in Brussels and discusses the possibilities of regional representation in the Council.

The Committee of the Regions and the Intergovernmental Conference of 1996 is the next issue discussed. Here emphasize is put on the Committee's own requests for an increase in status and power. However, the real power of the Committee will still be weak even if all its requests are granted.

Last but not least the paper addresses the future role of the regions in the European Union. The shift in relative power between the federal and regional level will create a demand for a stronger role of the regional level in the decision making process. If the European Union develops into a federal state it proposes the Council as a future Chamber of the States and Regions, sharing its power with the Commission and the European Parliament.The Committee of the Regions will then be a committee for advice and contact between the institutions of the European Union and the local authorities. If the Union do not develop into a federal state it supports the idea of promoting the Committee of the Regions as a second chamber.

The views expressed in this paper are those of the authors and cannot in any way be interpreted as the views of the European Institute of Public Administration.

2 Introduction

It is the purpose of this paper to make a small contribution to the on-going debate on the future of the regional level in the decision making process of the European Union.

The term region has been defined in many different ways, used in many connections and for many purposes. The purpose of this paper is to discuss the role of the regions in the governmental structure of the European Union and its Member States.This governmental structure can for most purposes be divided into five levels. The Union level, the Member State level, the regional level, the provincial or county level, and the city or local municipality level. Some of the Member States do not have a regional level. This is the case of not only small Member States, as Finland, Ireland and Luxembourg, but also of a large Member State as the United Kingdom. These states have today a unitary system. For the smaller states there is and will probably never be a need for a regional level, as we have defind it. However, for the larger states with a unitary system there is today a clear tendency for a stronger regional role. This is the case not only in Spain, where the power of the regional level has increased dramatically in the last years, but also in France where the regional level in the last years has gained considerable power. In the United Kingdom it is yet not clear if there is a move to towards more power for a regional level. Examples of Member States which today have a clear powerful regional level are Belgium with its regions and communities, Germany with its "Länder", Italy with its "Regioni" and Spain with its "Comunidades autónomas".

It is the evolution of the power and role of such regions in the governmental structure of the Union that will be the primary target for our discussion in this paper. However the development of the power of this regional level is strongly connected with the evolution of the regional policy of what today has become the European Union. The evolution of this policy and the content of this policy today will therefore be outlined in the next sections. The distinction between a regional policy and a policy towards or vis a vis the regions will then be discussed.

The establishment of the Committee of the Regions is one instrument in the Union policy towards the regions. The creation, functioning and future of this Committee together with other forms of regional participation in the decision making process of the European Union will therefore be discussed in the next sections. Finally the future role of the regions in the context of the Intergovernmental Conference 1996 and beyond will be addressed.

3 Evolution of EU Regional Policy

The second world war was a catastrophe that ended with a Europe in ruins. The prewar economic, social and political institutions of the European countries had more or less collapsed. Two world wars in 30 years and the threat of a new war created a climate for new thoughts and initiatives of how to integrate the former foes and friends in Europe. The purpose of all these initiatives was clearly to secure peace and create prosperity for all people in Europe.

The Treaty of Paris - No Need for Regional Policies. In 1951 six countries: Belgium, France, Germany, Italy, Luxembourg and the Netherlands signed what is called the Treaty of Paris. From July 1952 this treaty established the European Coal and Steel Community ECSC. It was the first time that member countries gave up national sovereignty to an overnational authority, although within a limited area. Their ambitions went further, however, towards a more extensive economic union.

From the preamble to the ECSC treaty it follows that the main purpose for the founders was to secure peace and that they were conscious of the fact that this only can be reached by concrete actions creating a real solidarity and by establishing a common basis for economic development. The mission of this Community was, according to Article 2, to contribute to the expansion of the economy, the development of employment and the improvement of the standard of living in the participating countries. Emphasize was put on establishing conditions that would assure the most rational distribution of production at the highest level of productivity. Although it also said that the continuity of employment should be safeguarded and that disturbances in the economies of the member States should be avoided, regional problems and regional policies are not mentioned at all.

The instrument of the ECSC community was the creation of an integrated common market for coal and steel. This was done by abolishing all customs barriers and other restrictions on the movement of coal and steel between the member states.

The establishment of the ECSC, together with the Marshall aid, and the general reduction of tariffs and removal of trade barriers, made possible by the work of GATT and OEEC fast created economic growth, rapid increase in employment and raising living standards in all the western European countries. However, this growth process also created technological changes, dramatic restructuring and as a consequence decline in employment in the primary and extractive industries which again caused extensive domestic migration.

All these formidable changes did not create any feeling of a need for regional policies either on the Community or on the national level during the 1950's. The predominantly domestic migration in these years did not seem to be of any major concern for the governments of the Member States. The reason was possibly that at that time the migration was seen and probably also felt, as a voluntary thing of no concern to the governments as long as housing was procured in the areas where people moved to get new employment.

The Treaty of Rome - The European Investment Bank. Parallel to the establishment of the ECSC a new initiative was taken for further European integration. This initiative and efforts led, after several setbacks to the establishment of the European Atomic Energy Community (EAEC) or EURATOM and the European Economic Community (EEC) by a treaty signed in Rome in 1957 and made operational from January 1st. 1958. The Treaty of Rome expresses very ambitious objectives. According to Article 2; "It shall be the aim of the Community, by establishing a Common Market and progressively approximating the economic policies of Member States, to promote throughout the Community a harmonious development of economic activities, a continuous and balanced expansion, an increased stability, an accelerated raising of the standard of living and closer relations between its Member States."

To be able to reach these objectives the Treaty facilitates:

1. The removal of barriers to trade among Member States.
2. The establishment of a single commercial policy towards non member countries.

3. The abolition of the obstacles to the free movements of persons, services and capital.
4. The inauguration of a common agricultural policy and a common transport policy.
5. The establishment of a system ensuring that competition shall not be distorted in the Common Market.
6. The application of procedures which shall make it possible to co-ordinate the economic policies of the Member States and to remedy disequilibria in their balance of payments.

In addition the Treaty created a European Investment Bank which should contribute to a balanced and smooth development of the Common Market in the interest of the Community. For this purpose the bank should also facilitate the financing of projects to develop less developed regions. This has been seen as the first step in the creation of EU regional policies.

The Common Market - Regional Tensions. The theory behind the Common Market was that the removal of trade barriers, tariffs and quotas, would increase efficiency in the allocation of resources and thereby free resources. This process would then lead to economic growth. Although it was understood that this growth would not be even, it was strongly believed that spillovers from the fastest growing regions would spread and generate growth also in the initially declining regions. This generated growth would in most cases be strong enough to solve the economic problems without specific measures. The economic integration that would follow would also lead to political cooperation and ultimately to a European Unity.

The 50's were years of high and sustained economic growth in most of the countries of Western Europe. The growth was particularly strong at the end of the 50's and the early 60's. According to most researchers this was caused by an enormous escalation of international trade. International trade grew in large part because of the removal of tariffs, quotas and other trade barriers initiated by the GATT and OEEC agreements. The first intra EC tariff reduction took place January 1st. 1959. Other rounds of reductions in tariffs and quotas among the six Member States followed with relative short intervals. The Community also started to erect a common tariff, CET, toward outside countries. This common tariff were then in the next years to be drastically reduced. Consequently it did not create a wall between the EEC and the countries outside. The first four years of the EEC customs union saw an increase in trade of more than 50% among the six member countries. According to Flynn (1966) this dramatic rise was, however, as much the result of "an increased activity of businessmen as of the actual reduction in tariffs. As soon as managers were convinced that the common market was going to be established they started to behave in many ways as if it was already in existence."

This rapid growth also led to continuous restructuring of the economies of the Member States in the 1960's. Technological changes led, for example, to a tremendous restructuring of the Coal industry. The labour force in coal mining in the Member States was reduced from more than 650 000 at the end of 1957 to less than 300 000 ten years later. The effects were a very high rate of unemployment and outmigration from the typical coal districts. However, this fact did not result in any action by the Community. The restructuring of the economy was still seen as a necessity to both the Community and to the national objectives of economic growth and raising living standards. Here it should be added that the Treaty of Paris in principle banned national assistance to the coal and steel industries. The Treaty of Rome was less restricted, national assistance could be introduced with the consent of the Commission if it went to areas with a very low standard of living or high unemployment or were aimed at promoting regional development.

However, the effects of the tremendous structural economic changes taking place started in the 1960's to be felt as regional problems in most countries of Western Europe. As a consequence many European States adopted regional development policies in these years.

But these regional policies, or this regionalism, was seen as what has later been called top-down regionalism. The growth process was regarded as tapping of underutilized resources in declining regions to increase national output. When this growth process became too strong it became desirable to create growth in the declining and often peripherary regions. This growth would then quell congestion in the growing mostly urban regions. It was seen as a zero-sum-game where all could win.The political justification was to enhance national solidarity and to secure support for the national policy from the peripherial regions

In the 1970's, according to Keating and Jones (1985), tension arose between the top-down perspective of the national policies and pressure arising from within regions. The pressure from within was caused both by growing resistance to change and by the fact that new social forces emerged. This tension increased with the launching of the programme for the creation of the internal market.

The Internal Market - The Need for Regional Policies. As early as 1970 a memorandum on industrial policy from the Commission said that in spite of the undeniable liberalisation of trade between the Member States, created by the Treaty of Rome, their six markets together were far from resembling a single market. The proof lay in the differences in the price-level for similar products between Member States. The Commission attributed the inflexibility of prices in part to

- differences in mentalities and tradition,
- inadequate knowledge of partners' products,

- routine loyalty to habitual products,
- the lack of adaption of a country's products to the tastes and habits of consumers in others,
- various strategies adopted by producers and distributors from the Member States.

To a large extent, however, the inflexibility of prices stemmed from the compartmentalisation of markets owing to the old and new barriers erected by the national authorities to replace the customs barriers. These barriers the Commission had even as early as 1970 started to wage a war against.

Fifteen years later, in 1985, the commissioner for the common market Lord Cockfield had drafted a White Paper on the completion of the internal market in which it clearly said that the "common" market was far from resembling a single market for a number of reasons. The main reasons were:

- customs delays and the administrative charges weighing on intra-Community trade;
- the differences between plant-health and veterinary standards, necessitating checks at borders; the formalities carried out at borders for statistical purposes;
- national differences in technical rules and standards for industrial products,
- the lack of openness of public contracts to foreign suppliers,
- restrictions on the freedom to provide certain services, in particular financial services, and on freedom of establishment with regard to certain activities, differences between the rates of VAT and excise duties, governed by the "rule of destination" and which had to be adjusted at the border of the Member States of destination,
- the application of monetary compensatory amounts to inter-Community trade in certain agricultural products in accordance with the rules of the common agricultural policy, and
- road transport licences and the checks on the conformity of vehicles with the various national regulations.

The White Paper also tried to compute the cost for the Community of not having created an internal market, the cost of Non-Europe, and came up with a programme to complete such an internal market by the end of 1992.

This programme for the elimination of all physical, technical and fiscal obstacles within seven years was in particular characterized by a more flexible approach to harmonization methods, the deliberate rejection of all easy ways out, which would in practice result in maintaining controls on trade at the internal borders, the adoption of a precise and binding timetable, which was indispensable to the success of the programme, the importance of which lay in its overall approach, and in the balance of interests involved.

The White Paper therefore listed a battery of almost 300 legislative texts, together with a timetable for their implementation, aimed at removing barriers to trade.

The new president of the Commission Jaques Delors, devoted his first period as president to the implementation of this programme.

The European Council adopted the Single European Act with its programme for the completion of the Internal Market at the Milan Summit in June 1985. This Act, which entered into force on July 1st. 1987 amended the EEC Treaty, confirmed the objective of completing an area without frontiers by 1992, and the timetable set out in the 1985 White Paper. It adapted the Community's decision making procedures and increased the scope for qualified majority as opposed to unanimous voting in the Council.

The objective of the completion of the Internal Market was to;
- merge the 12 national markets to form one single enlarged market,
- create an expanding dynamic market,
- ensure that this market was flexible enough to channel human, material and financial resources to those areas where they would be best used.

The Internal Market is both narrower and deeper than the Common Market of the Treaty of Rome. It is narrower in the sense that it does not embrace the various common policies, in particular the Common Agricultural Policy (CAP). On the other hand it goes deeper since it assumes the elimination of obstacles which were lawful in Community law and it also aim at positive integration, implying that account must be taken of objectives of economic, social and legislative cohesion.

When the internal market programme was launched it was clearly understood that the economic forces of such a single market would have unwanted effects in many areas. Furthermore it was a common realization that the Community ought to take a stronger lead in the creation of programmes that would counter-attack the unwanted effects of these forces.

4 Regional Policy of the European Union

The Beginning. A formal European Community regional policy was not introduced until the mid 70's, when the European Regional Development Fund, the ERDF, was created in 1975. This was, however, a small fund. In the first years its share of the Community budget was well under 5%. Its regulations, with for example quotas of its expenditures for Member States, prevented it from being a very effective instrument.

According to Armstrong (1995), the EU regional policy took almost 15 years to come of age. Until a set of massive reforms in 1988, regional policy struggled to establish itself. The EU budget was dominated by agricultural spending and the Member States considered regional policy a national matter.

However, with the launch of the Single Market programme it became clear that more assistance would have to be given to regions that quickly had to restructure their economies.

Objectives and Instruments. The 1988 reforms establish clear objectives for the three structural funds, the ERDF, the European Social Fund and the Guidance Section of the Agriculture Fund.

Five priority objectives for these Structural Funds were established. Objective 1, 2 and 5b are specifically regional in nature; they involve measures restricted to certain eligible regions or parts of regions. Objective 1: development and strucural adjustment of the regions whose development is lagging behind. Objective 2: converting the regions or parts of regions seriously affected by industrial decline. Objective 5b: development of rural areas. Objective 5 were created with a view to the reform of the common agricultural policy.

Objectives 3, 4 and 5a on the other hand cover the whole of the Community. They are concerned with combating long term unemployment, occupational integration of young people and speeding up the adjustment of agricultural structures.

In real terms the budget for these funds, assisting the disadvantaged regions to respond to the challenges of the Internal Market, were doubled between 1988 and 1993.

The reforms of 1988 also established the principles on which the EU regional policy rests:

1. Concentration of assistance in the areas of greatest needs.
2. Careful coordination and planning of EU's attack on regional problems.
3. Partnership between the Commission, Member State governments, regional and local authorities and organizations.
4. Additionality with EU funds complementing the regional policies of the individual Member States.
5. A focus on programmes of assistance rather than piecemeal help for individual investment project.

According to Armstrong (1995) the reforms of 1988 have served the EU well. The new system attracted very little criticism. These reforms also stimulated regional and local development initiatives.

Dramatic Budget Increases. In 1993 some relative small changes took place in the formulation of policy objectives. These changes were designed to cover the budget years 1994-1999. The budget for what can be termed structural actions for these years were, however, largely increased compared to previous years. The budget for 1999 is in real terms 50% higher than in 1994. The importance attached to structural policies in the Community in the years ahead can best be seen in the share of Community expenditure on structural actions, that is the Structural Funds and the Cohesion Fund. In 1993, the last year of the old programming period, these expenditures accounted for 31% of the total. In comparison 51% of total expenditure went for agriculture. By 1999 the share for structural measures will have risen to 36% while the share for agriculture will have declined to 46%.

Here one should also keep in mind that the expansion of EU regional policy expenditure has come at a time when Member States budgets have had to be reduced. Many regions with severe structural problems have therefore come to look to Brussels for support instead of to their own national governments.

In the revised 1993 regulations for the Structural Funds provisions were made to include areas which are suffering from a decline in fishing activity in regions or areas eligible under the regional Objectives. Furthermore a new Objective 3 combined the tasks of the previous Objective 3 and 4 but also aimed at "facilitating the integration ...of those threatened with the exclusion from the labour market". A new Objective 4 took care of the new tasks laid down for the Structural Funds in the Maastricht Treaty: "to facilitate workers' adaptionn to industrial changes and to changes in production systems." Objective 5a was extended to include aid to modernize and restructure fisheries. These changes also included a commitment to *additionality* by preventing Member States from substituting EU regional policy expenditure from their own spending.

Completion of the accession negotiations with Austria, Finland, Sweden and Norway, and the ultimate accession of the first three countries into the European Union, also included a further change, Objective 6 regions. These type of regions were designed to accommodate the particular difficulties in the sparsely populated, climatically extreme northern regions of the Nordic countries..

Effects of the Regional Policy. European economic integration, through the establishment of the customs union and the internal market, which led to a total restructuring of many regional economies and the development of the regional policies of the European Community, have had a profound effect on the regions of Europe. It has caused an awakening of regional economic and cultural identity in many regions. This awakening has not only come in regions which directly have benefited from this policy but also in many regions which have not received any aid and which are not candidates for receiving any in the future.

The economic and political integration which has occurred among the Member States has also made it more "comme il faut" for the citizens of a particular country to identify with a particular region of that country. This identification has also created a process in which regional and local authorities have demanded more participation, more independence and more self governing. As a consequence we have seen that the regional and local level has been strengthened in many of the Member States the last decade. This process can also be said to be in line with Article A of the Maastricht Treaty; "This Treaty marks a new stage in the process of creating an ever closer union among the peoples of Europe, in which decisions are taken as closely as possible to each citizen."

5 Regional Policy or a Policy of the Regions

The regional policy of the Community can, however, hardly be called a policy towards the regions. It is therefore necessary to distinguish between regional policy and EU policy vis-a-vis the regions.The first is a policy to assist regions which for different reasons have an economic disadvantage compared to other regions. This kind of policy is also often called development policy. The other is a policy geared at drawing all regions and local authorities, regardless of their economic, geographical or political standing, closer into an integrated political decision-making process of the European Union.

Although most of the Internal Market legislation is now in place in the Member States the implementation of the legislation still has some way to go. Furthermore the full economic effects of the Internal Market will take a long time to be fulfilled. This goes both for the positive and the negative effects of the restructuring that will follow from the dynamic changes now taking place.The full regional effects of the Internal Market is yet unknown. It will of course also depend on the industrial structure of the individual regions.

The Community will in the next years take further steps towards a full economic integration. Stage 2 of the process leading to the economic and monetary union, EMU, has already been taken while the final stage, a single currency, will according to plans be introduced in 1999. It is generally agreed among the economists that a single currency will help regional development in as far as it reduces transaction costs and eliminate exchange rate risk. At the same time, however, the Member States will lose certain fiscal and monetary policy options as well as their ability to adjust their exchange rates. Devaluation has for example been used by countries to offset a loss in international competitiveness. But most economists also agree that devaluation in itself does not solve the underlying causes for the loss in competitiveness. In a single currency system Member States will also have to adopt a tighter fiscal policy, which might mean less money for regional policies in the short run. The revised regulation of the structural funds from 1993 recognizes the higher need for Community intervention to counteract this effect. It is also recognized that the tighter fiscal policy hopefully will prevent macroeconomic imbalances which in turn will create conditions for faster economic growth. Currencies from Member States or countries outside the Single currency area will also be more vulnerable to speculations. This will require an even tighter monetary and fiscal policy from these Member States or countries. Economists therefore agree that a delay of entry into the single currency system is undesirable for the weaker Member States.

The political will to take the step towards an economic and monetary union is also a manifestation of the move to "an ever closer union" and a greater sharing of decision making between the union and the national-, regional- and local level.

6 The Committee of the Regions

To meet the recognized need of integrating the regional and local authorities into the decision making process of the Community, the Treaty of the Union or the Maastricht Treaty established in its Chapter 4*, Article 198a to 198c, a committee with advisory status consisting of representatives of regional and local bodies, the Committee of the Regions, COR.

The Creation - The Three Alternatives. The history of the COR dates at least back to 1988 when the Commission made the decision to create a consultative Council of Regional and Local Authorities. (Dec. 88/487 OJ L 247723). This Regional Council should be consulted by the Commission on every question having an impact on regional development and on every legislative initiative linked to the regional policy of the European Union.

On the governmental level it was the Federal Republic of Germany that took on the role as speaker of regional interest in the Community. This is of course due to the wide ranging authorities and competence of the German "Länder". Germany received strong backing from other federal states, first from Belgium and Italy and later from Spain. The more centralized Member States were less enthusiastic to strengthen the role of these subnational bodies. They also feared that it would complicate the decision-making process in the Community unnecessarily. From the very beginning of the Intergovernmental Conference on the Political Union in 1990, the IGC 1990, it was obvious that the German, Belgian and Italian initiative in favour of the regions would result in the creation of a body representing them.

The question was to decide what status such an entity should have and to what extent it should be involved in deliberations at Community level.

Three possibilities were in the end discussed:

1. A consultative body attached to the Commission.
   The Commission's proposal was simply to reinforce the Consultative Council of Regional and Local Authorities mentioned above. This proposal was rejected by the German, Belgian and Italian delegations who did not want to subordinate the future of this new entity to the Commission.

2. An entirely independent body.
   The German delegation favoured an entirely independent body. But the majority of the other delegations rejected this for concern over complicate decisions and for budgetary reasons.

3. A body attached to the Economic and Social Committee, ESC.
   This proposal to enlarge the recruitment of the members of the ESC by including representatives of the regions was immediately rejected by the ESC itself. It pointed out that a modification of its structure would jeopardize the political balance in the ESC, which provides institutional representation for the various categories of economic and social activities. Instead the ESC proposed to create a regional entity which should use the general secretariat and the infrastructure of the ESC.

The Draft Treaty on the Union put forward by the IGC on June 18th 1991 proposed a Committee of the Regions "attached to the Economic and Social Committee". As this formulation seemed to subordinate the Committee of the Regions to the ESC, the German, Belgian and Italian delegations insisted to cancel the words "attached to the ESC". In the final text of the European Union Treaty, otherwise very close to the draft, these words were taken out from Article 198 a.

The Committee of the Regions, COR, is entirely independent from the ESC in its statutes, competence, and consultation, but shares with it its organizational structure and institutions (ref. Protocol no. 16 on the Economic and Social Committee and the Committee of the Regions annex to the EUT)

The Advisory Function. The Committee can become involved in the decisions of the European Union in three different ways:

1. Obligatory involvement in cases provided for in the Treaty. This cases are:
   - Education, vocational training and youth. (Art. 126.4 ECT)
   - Culture (Art. 128.5 ECT)

- Public health (Art. 129.4 ECT)
- Trans-European networks (Art. 129 d ECT)
- Economic and social cohesion (Title XIV ECT)

2. Involvement through the Council or the Commission

3. Involvement acting on its own initiative
   - Within the framework of consultation with the Economic and Social Committee, where specific regional interest are involved.
   - With respect to all Community policies, where such action is considered appropriate.

Representation in the Committee. The COR consists of 222 members from the 15 Member States and an equal number of alternate members, representing regional and local authorities. The number of representatives from each Member State is the same as for the Economic and Social Committee. Table 1 shows the number of COR representatives from each Member State, together with total population in each country and population per representative.

Tab. 1 Representation in the Committee of the Regions

	Rep.	Pop. 1992 mill.	Pop. mill/ rep.
Austria	12	7.9	0.7
Belgium	12	10.1	0.8
Denmark	9	5.2	0.6
Finland	9	5.1	0.6
France	24	57.6	2.4
Germany	24	80.1	3.3
Greece	12	10.3	0.9
Ireland	9	3.6	0.4
Italy	24	56.9	2.4
Luxembourg	6	0.4	0.1
Netherlands	12	15.2	1.3
Portugal	12	9.9	0.8
Spain	21	39.1	1.7
Sweden	12	8.7	0.7
United Kingdom	24	57.9	2.4
Total	222	368.0	1.7

Source: Eurostat 1994.

Table 1 makes it possible to compare population and representation for each of the Member States. Such a comparison shows that the small countries have a larger share of the representatives than they would have with a proportional distribution among the Member States. Luxembourg, for example, has less than 100.000 inhabitants behind each of its six representatives, compared to Germany who has 3.3 million inhabitants behind each of its 24 representatives. This representation, which is very favourable to the countries with a small population and unfavourable to the countries with a large population, may constitute a problem in itself since the committee is supposed to represent the regional and local level in all the Member States.

Do the Committee in practice represent both the regional and local level? This question is hard to answer. An examination of the affiliation of the representatives show that more than three quarter of the representatives from the Member State with a distinct regional level comes from the this level. In Belgium all 12 representatives comes from either the regions or the communities not one from the counties or local municipalities. The need for a regional representation in Brussels must therefore have been felt more urgent than having representatives from the counties or local municipalities.

The Constituent Bodies. The constituent bodies of the COR are: The *Plenary Session*, in which all 222 members of the committee meet. The members and the equal number of alternate members are appointed by the Council for a period of four years. The Plenary Session held its first meeting in the Spring of 1994.The *Bureau* is elected for 2 years. It compromises the Chairman, the first Vice Chairman and 34 other members.The task of the Bureau is to prepare and organize and coordinate the work of the Plenary Session and expert committees. The Bureau also process the opinions of the COR by transmitting a report to the Council and Commission which the President of the COR has presented to the Plenary Session and which it has approved. The COR has eight regular sub committees or expert committees for special fields called commissions and one Special Commission on Institutional Affairs.

Opinions Issued by the Committee. In its two years of functioning the Committee has given its opinion on a wide range of issues. At its second session in the Spring of last year it issued its first twelve opinions. The most important being on the Councils proposal for establishing a Cohesion Fund. The Committee broadly endorsed the proposal but recommended that the competent authorities at regional and local levels should be closely involved in the management of the Fund. In the same session the Committee called for a better coordination of Community initiatives under the structural funds. It also expressed its opinion on several aspects of the Trans-European Networks.

In the autumn sessions of last year the Committee criticized the Commission's White Paper on Growth, Competitiveness and Employment, for among other things its lack of sectoral, local and regional analysis and its cursory treatment of the role that regional and local authorities can play in creating growth and employment. In these sessions the Committee also expressed its opinions on further aspects of the Trans-European Network stressing the need for cooperation with regional and local authorities. The Committee's opinion on the Commissions proposal for a directive on the right to vote and stand in municipal elections is important. It stressed that derogerations to the principle of non-discrimination on the basis of nationality, which Member States chose to apply, should be transitional and defined as narrowly as possible. The Committee also said that, as a follow-up to the current proposal, the European Union should address the civic rights and duties of third country nationals legally resident in the Union.

The Committee made a strong resolution on the principle of subsidiarity in which it called for this principle to be affirmed in the revision of the Maastricht Treaty and applied to all institutional levels: European institutions and bodies, the Member States, regions and local authorities. It also asked to be granted a bigger role in the decision making process and to be accorded the right to bring proceedings before the Court of Justice regarding breaches of the principle of subsidiarity affecting the powers of local and regional authorities.

It is only two years since the Committee has started its work. It might therefore be to early to ask the question if the results are such that they make all the efforts and work done by the members of the Committee have been worthwhile. Although it is too early to make a final judgement we find it hard to see that the time and work spent by the members of the Committee can be justified by its achievements.

7 Other Forms of Regional Participation

Pressure Groups in Brussels. The Committee of the Regions is not the only form of introducing regional interest in the policy-making of the European Union. In the institutional architecture of the EU there are also other possibilities. In Brussels one will find all kinds of "pressure-groups" lobbying the interest of their Region. There are for example an important number of, what is known as, "Information Offices" whose number has increased dramatically the last few years. The legal status of these offices is not clear, but it is accepted that they engage in information procession and lobbying activities. Besides these forms of regional participation there is also the important aspect of the internal forms of regional participation on policy making within the Member State itself.

Regional Representation in the Council. The relation of the Regions with the European Community is a wide field and in the context of this study it will only be focused on the internal policy making of the Member States concerning EU matters and the representation of regional ministers in the Council. The strongest legally regulated forms of regional participation can be found in Belgium, Germany and Italy. In Spain and France regional matters and the question of regional representation have gained increasing support in the last few years. However it is still a long way to go before the regions will have the same status as in the first mentioned countries. Some Member States do not have any regional participation, namely Greece, which is one of the most centralised States in the Union, Luxembourg, which has no need since the State itself is a region in the EU context, Denmark, Sweden, Finland, United Kingdom Portugal and Ireland.

The Belgian Model. One of the Member States with regional participation deserves a closer look. Belgium brought up the discussion on the representation of the Member States in the Council. This led to the change of Article 146. Before the change each Member State had to be represented by a member of the national government, that is the federal government in the case where the State has a federal system. This regulation did not respect the internal distribution of competence between the Regions and the central government in States with a regional structure. For the case of Belgium some fields of competence, like education and culture were at the exclusive legislative competence of the regional level. Belgium therefore wanted to change Article 146 ECT in order to make it possible for a regional Minister to represent the Member State in certain fields of competence. The question of fields of competence and the elaboration of a position on the national level has so far been left entirely to each Member State. In Belgium an informal agreement, within the framework of the Interministerial Conference, between the Federal Government, the Regions and Communities has been made about representation in the Council.

The Belgian participation in the internal decision-making process besides representation in EU matters is based on the Special Law on Institutional Reforms of 08.08 1980, 08.08 1988 and the changes by the Law of 05.05 1993 in the International Relations of Regions and Communities.

The German Case. Germany introduced a new article in its Grundgesetz by a Law of 21 December 1992. Article 23, the so-called "European Article" regulates in its paragraphs 4 to 6 the participation of the Länder in EU matters. It is a strongly structured system differentiating three levels of participation. The weakest form of participation is the consideration of the opinion of the Bundesrat in the fields of exclusive competence of the federal government as far as interests of the Länder are concerned, and in the fields in which the federal government has legislative power apart from above mentioned exclusive competence. The Bundesrat is the representative chamber of the Länder, consisting of members of each Land's government. The strongest form of participation is the representation of the Federal State by a representative of the Land (ref. Art. 146 ECT); this is the case when mainly the exclusive competence of the Länder are concerned.

The Spanish Case. Spain, although a Member State with a strong regional structure, has only few mechanisms for regional participation. There is yet no possibility for a regional representative to represent the federal State in the Council. The only form of institutionalised regional impact is the "Sector Conference on EU matters". This working group co-ordinates the information between the Central State and the "Comunidades".

8 The Future of the Committee of the Regions

The Intergovernmental Conference 1996. The Treaty of Maastricht provides in Article N for a conference in 1996 of representatives of the governments of the Member States "to examine the provisions of the Treaty for which revision is provided, in accordance with the objectives set out in Articles A and B." Paragraph 2 of Article A reads: "This Treaty marks a new stage in the process of creating an ever closer union among the people of Europe, in which decisions are taken as closely as possible to its citizen".

Here it should also be mentioned that the ratification of the Maastricht treaty proved difficult and that, according to Ludlow and Ersbøll (1995), the so-called post-Maastricht ratification crises added a new dimension to the debate about the IGC 96. This new dimension consists of a greater sense of urgency in the debate about the "democratic deficit", the need for greater transparency and a keener awareness of the actual and potential importance of the principle of subsidiarity.

The IGC 1996 and the COR. Already there has been a lot of discussion among the members of the Committee of the Regions about the future of the Committee. The most controversial issue has been the organization of the Committee itself and its place within the European institutional architecture.

In September 1994 the Committee of the Regions set up a special Commission on Institutional Affairs to draw up proposals for the study group responsible for making preparations for the Intergovernmental Conference in 1996 who will revise the Treaty of the Union.. The discussion of this commission focused on the following three aspects of the Maastricht Treaty:

- The principle of subsidiarity
- The access of the COR to the Court of Justice for actions against legal acts of the community institutions
- The structure and the functions of the COR itself.

The Commission appointed a special committee headed by Jordi Pujol, President of the Region of Catalonia, to draw up a report on the basis of these three aspects.

The Requests of the Committee. In April this year, according to the press release, the Commission on Institutional Affairs, on behalf of the Committee of the Regions, adopted almost unanimously, after a long and somewhat difficult debate, the opinion of the Jordi Pujol report on the revision of the Maastricht Treaty. The main requests in the report are:

1. *On the principle of subsidiarity.* Article 3b of the Treaty should, according to the Committee, be reworded stating that the Community should intervene "only if and in so far as the objectives of the proposed action cannot be sufficiently achieved by Member States, or by the regional and local authorities endowed with powers under the domestic legislation of the Member State in question". The Committee also wants "the respective powers of the Union and Member States to be clearly defined".

The Committee proposes to modify the third paragraph of Article 173 of the Treaty so that the Court of Justice "shall have the jurisdiction in actions brought by the Committee of the Regions against violations of the principle of subsidiarity" as well as "in actions brought by the regions whose legislative powers may be affected by a regulation, directive or decision".

2. *The place of the Committee in the institutional architecture.* The Committee of the Regions should be made an institution alongside the European Parliament, Council, Commission, and Court of Auditors. This implies and amendment to Article 4 of the Treaty.

The Committee of the Regions should also have the right to make appeals for failure to act before the Court of Justice. This implies that if the Committee should not accede to the rank of institution a modification has to be made to the first paragraph of Article 175 of the Treaty. This paragraph would then be worded: "Should the European Parliament, the Council or the Commission, in infringement of the Treaty, fail to act, the Member States, the other institutions of the Community and the Committee of the Regions may bring an action before the Court of Justice to the infringement established":

3. *The right of initiative.* The Committee of the Regions requests to be able to "cooperate with the Commission when the latter take initiatives". It therefore proposes an additional paragraph to Article 198C of the Treaty. This paragraph stipu-

lates that the Committee shall offer the Commission its cooperation and advice in drawing up legislative programmes and Green and White Papers, and in preparing other initiatives in respect of the policies affecting the powers of regional and local authorities".

4. *The consultative function.* The Committee of the Regions requests a strengthening of its consultative function and it should explicitly be provided for in cases where consultation is provided for of the Economic and Social Committee, for development cooperation policy, and for Union citizenship and state aid.

5. *The political mandate, and legitimacy of the Committee's members.* The Committee of the Regions requests that first paragraph of Article 198 A of the Treaty should be amended to read as follows: " A Committee, hereinafter referred to as the Committee of the Regions, consisting of representatives appointed on the recommendation of regional and local bodies, who have their mandate from the electorate or are politically accountable to an Assembly elected by universal suffrage, is hereby established with advisory status".

6. *Administrative autonomy.* The Committee of the Regions requests "full organizational and budgetary autonomy vis-a-vis the Economic and Social Committee".

7. *The principle of local autonomy.* The Committee of the Regions also wants the *principle of local autonomy*, as defined in the Council of Europe's charter of Local Autonomy, to be included in the Treaty. Furthermore, it wants a better Community cooperation in the fields of justice and home affairs as well as in developing the concept of European citizenship. It therefore wants a catalogue of fundamental rights to be included in the Treaty and that the Committee of the Regions should be involved more explicitly in action in these areas.

9 The Future Role of the Regions

Beyond the IGC 1996. It is quite clear that if all the requests of the Committee of the Regions should be accepted at the Intergovernmental Conference 1996 it would increase the power of the Committee from a purely advisory committee to an institution with the right to exercise some power. However, the role of the Regions, in the decision making process of the European Union, will still be rather limited. This was also confirmed by Jaques Santer, President of the European Commission, when he in a speach to a plenary meeting of the Committee commented on its proposals on the revison on the EU Treaty, according to "Europe"(No 6569), said: "You have opted for reinforcement of the advisory role and not for a co-legislator role", and he added, "It is a wise and realistic option".

It is probably true that it is wise at this stage, for the Committee, not to go further. The Committee only got started in the Spring of 1994. However it is also clear that many members of the Committee, and many regions which they represents are not satisfied with the influence that the regions have in the decision making process of the European Union. Here it is also worthwhile to remind ourselves of the fact that the inhabitants of many regions in Europe feel a very strong identity with their particular region and that many of the regions have a population that far exceeds many Member States. It should also be kept in mind that several regions are historic nation-states. The debate of the role of the Regions in the future institutional structure of the European Union will therefore continue beyond the Intergovernmental Conference of 1996.

A Shift in Power. The main objectives of the European Union are outlined in Article B of the Treaty of Maastricht. Here it says clearly that the objectives are

- to promote a balanced and sustainable economic and social progress through creation of an area without internal frontiers, through the strengthening of economic and social cohesion and through the establishment of an economic and monetary union, ultimately including a single currency;
- to assert its identity in the international scene, in particular through the implementation of a common foreign and security policy including the eventual framing of a common defence policy, which might in time lead to a common defence;
- to strengthen the protection of the rights and interests of the nationals of its Member States through the introduction of a citizenship of the Union
- to develop close cooperation on justice and home affairs

These objectives shall be achieved while respecting the principle of subsidiarity.

When all these objectives will have been reached a substantial shift in the exercise of power in the areas of monetary and financial policy, foreign policy, and defence policy, from the governments of each Member State to the institutions of the Union will have taken place. In the Member States with a strong regional level, as in the case of Belgium, Germany, Italy and Spain, this can also be interpreted as a *relative* shift in power from the Federal government to the regional level. In the Belgian case from the Federal level to the Regions and Communities and in the Germany case from the Federal level to the Länder.

A Stronger Regional Role. In the last decade we have seen a shift in the exercise of power from the federal to the regional level in many of the Member States. This has clearly been the case of Belgium, Italy and Spain. The power of the regional level seems also to have increased in France, allthough, according to Balme (1995), their competences and finacial powers remain week. In the United Kingdom the question of regional governments is still unsolved.

The absolute and relative shift that has taken place in the exercise of power between the federal and regional level within many of the Member States has increased the demand for a stronger representation of the regional level in the architecture of the European Union. The establishment of the Committee of the Regions is one result of this increased demand.

There is absolutely no sign that the increased demand for more regional power will fall off in the years to come. On the contrary the preasure for increased implementation of the principle of subsidiarity is felt not only by the institutions of the European Union but also in each Member State.

Parallel with this domestic shift in each Member State there also seems to have been a shift in power from both the Commission and the Council of the Union to the European Parliament. A Parliament which defacto power was substantially increased when direct elections were introduced. The decision making procedures of the Union has to be viewed as a dynamic process. A process in which the European Parliament over the last years has gained substantially more power. It seems to us that the Parliament already has gained power that goes beyond what is written into the Maastricht Treaty. Conclusions about the constitutional system of the European Union drawn from only reading the treaties may therefore lead us astray. It will therefore also be the work of the IGC 1996 to write into the treaties the development that has already taken place. The political, economical and institutional system of the European Union has to be viewed as an ongoing process. A process that so far has moved in the direction of more democracy. This process has also made the Parliament and the other institutions of the Union more interesting and more important viewed from the regional level.

The important question is how this strengthening of the regional role can be achieved.

The Council as a Chamber of States and Regions? From our discussion it is clear that although the role of the regions in the institutional architecture of the European Union has been strengthened in the last years their role remains limited. It is also clear that a strengthening of the Committee of the Regions, as requested by the Committee itself, will not give the regions a substantial role in the decision making process of the Union. Furthermore it is clear that the preasure for a real place for the regions in the decision making process of the European Union is growing.

The important question is than how, in the institutional structure of European Union, the regions can be given a role in the decision making process? The answer to this fundamental question will depend on the future governing structure of the Union. If it develops into a true federal state then we ought to start a discussion into the possibilities of promoting the Council as a second chamber in the Constitution of the European Union. Such a discussion will go beyond the Intergovernmental Conference of 1996 and into the next century.

This 'new' Council could be a Chamber of States and Regions. It could then be a modus vivendi, in which the interests of regions and nations were integrated into a greater whole. If such a 'new' Council is created the balance of power between this 'new' Council and the Commission and the Parliament would, as a consequence, have to be redrawn.

The 'new' Council could have a total representation that would not largely exceed the total voting power of the present Council.The regions of each Member State should have a direct representation in this new Council together with the nation state level. Representation should also try to be more in accordance with population in each region and each Member State. A minimum representation for the smaller Member States should, however, be built in.

The Committee of the Regions will then be a committee for advice and contact between the institutions of the European Union and the local authorities, that is the provinces, counties and local municipalities.

If the European Union do not develop into a federal state the Council, representing the governments of the Member States, will also in the future have a role as a motor in the governing structure. In such a case the old idea, advocated by some members of the Assemby of European Regions (AER), to promote the COR as a second chamber should be seriously discussed.

References

Armstrong, Harvey W. (1995), Crisis or Opportunity? Regional Policy and Structural Aid, The European Union and the 1996 IGC, A Discussion paper of the Jean Monnet Group of Experts, Center for European Union Studies, University of Hull, England.

Balme, Richard (1995), French Regionalization and European Integration: Territorial Adaptation and Change in a Unitary State, In Jones and Keating [Eds.] (1995).

European Commission (1994), Competitiveness an d cohesion: trends of the regions, Fifth periodic report on the social and economic situation and development of the regions in the community, Office for Official Publications of the European Communities, Luxembourg.

European Commission (1995), Intergovernmental Conference 1996, Commission Report for the Reflection Group, Office for Official Publications of the European Communities, Luxembourg

Flynn, Edwin Harvey (1966), The European Common Market: Its production, trade, and welfare effects, Microfilms International, University of Michigan, Ann Arbor.

Keating, Michael and Barry Jones [Eds.] (1985), Regions in The European Community, Clarendon Press Oxford.

Keating, Michael and Barry Jones [Eds.] (1995), The European Union and the Regions, Clarendon Press Oxford.

Ludlow, Peter and Niels Ersbøll (1995), Towards 1996: The Agenda of the Intergovernmental Conference, Center for European policy Studies, Brussels

Morphet, Janice (1994), The Committee of the Regions, Local Government Policy Making, Vol. 20, Number 5 May 1994.

Neven, Damien (1995), Regional Convergence in the European Community, Journal of Common Market Studies, Volume 33, No. 1.

O'Neill, Normann (1995), Crisis or Opportunity? Regionalism and Local Government, The European Union and the 1996 IGC, A Discussion paper of the Jean Monnet Group of Experts, Centre for European Union Studies, University of Hull, England.

Office for Official Publications of the European Communities (1993), European Union - Selected instruments taken from the Treaties, Book I Volume I, Brussels-Luxembourg

Regional Policy and Federalism in the Process of International Integration

Riccardo Cappellin
University of Rome „Tor Vergata"
Department of Economics
Via di Tor Vergata
I - 00133 Roma

1 Introduction

Regional policy, both at the EU and at the national level, is confronting serious challenges, which may question even its existence in the medium and long term.

In fact, the process of globalisation of markets and the existence of large multinational firms may make irrelevant an analysis of the economy within a limited regional framework and it could vanify the capabilities of governments to affect the long term evolution of the regional economies.

The increasing consensus on the value of market competition and the opposition to policy measures, which may distort market prices, have lead to a general dismissal of the dirigist approach of macroeconomic or macrosectoral, comprehensive and highly formalised economic planning, but may endanger also the concept of regional economic and territorial policy, which still represents a major field of activity of national and European institutions.

Finally, the increasing popularity of the regionalisation movements and claim of federalist reforms in most European countries may make increasingly politically difficult for national and EU authorities to support a coherent policy aiming to a greater cohesion between developed and economic lagging regions.

This paper aims to indicate that the network paradigm represents and useful analytical approach in the interpretation of these three related problems. In particular, the paper first investigates the process of internationalisation of regional economies. Then, it indicates the characteristics of a new "market oriented" economic policy at the regional level and, finally, it analyses the scope for national regional policies and for vertical cooperation between the State and the Regions within a federal system.

2 The Internationalisation Process of Firms

Firms are compelled to deep organisational transformations in their international structure of marketing, of production and of technological development of their products to exploit the opportunities of global markets. The process of reorganisation of firms is tightly related with their expansion at the international level. The internationalisation of firms requires a high decentralisation of the operative functions and the creation of flexible alliances with foreign firms.

In the past, the concept of globalization of markets indicated increasing flows of products and services at the international level and th existence of a price determining mechanism operating at the international level and not at the national level, such as with the markets of foreign currencies, bonds and stocks.

Nowadays, the concept of the internationalisation process indicates that also the behaviour of the individual economic actors is changing and becoming transnational. This implies an overall international integration not only of the markets of the products, but also of the supply side of the national economies, represented by the firms, the trade unions, the public institutions, the research organisations, etc. In fact, the internationalisation process has a clear social and institutional dimension, as it implies a change in the institutional and social mechanisms, which links the various economic and social actors not only in an international framework, but also within each national or regional economy (Bianchi P., 1995).

The internationalisation of economies is a broader process than that of the globalization of markets and it does not imply an increasing homogeneity within a "global village". It implies instead an increasing interdependence and the need for a policy of differentiation or of exploitation of the differences characterising each actor and of the original role which he may perform (The Economist, 1993 and 1995; Ohmae, 1990).

These changes may be defined as the shift of the industrial system of the advanced economies from an "international" structure to an "interregional" structure, as the economies of the various countries become integrated in similar way as those of the regional economies within the same country (Krugman, 1991).

However, the over ambitious aim to rediscover economic geography of Krugman demonstrates that the concept of increasing return is not the well known "chicken which makes golden eggs" and that the analytical results, which may be derived, such an higher geographical concentration, are in fact contained in the hypothesis themselves. On the contrary, the crucial characteristic of space is not concentration, as it is typical of the spaceless world of international economics, but geographical dispersal and differences and specialisation among the various industrial or urban poles of the territory: issues which have been the object of research in regional economics, at least in the last two decades.

In fact, "ever since Adam Smith economists have recognised that ...specialisation and division of labour have made possible the improved productivity that arises from technological change, better resource allocation, and specialised production, the key underlining features of modern economics" (North, 1987).

In a world of increasing specialisation of each production unit the crucial analytical tools can not be an aggregate production function and economies of scale, but rather transaction costs and network economies (Williamson, 1975; Coase, 1992), which may be applied also in a geographical or territorial perspective to explain economies of agglomeration in urban centres (Cappellin, 1988) or the process of industrial diffusion outside of old industrial regions, as indicated by the case of industrial districts of small and medium size firms in various intermediate regions (Becattini, 1990; Scott and Storper, 1990).

These new analytical paradigms, not only innovate those of horthodox microeconomics, but they may also be transferred from the analysis of individual sectors and regions to an international context. In that perspective, the recent changes in the organisation of large multinational firms emphasize a trend of decentralisation and of increasing regional embeddedness.

In fact, within the traditional model the large multinational groups were organised through subsidiaries operating in the individual countries and producing and distributing all the products of the group. Thus, these foreign subsidiaries represented a smaller copy of the mother firms.

On the contrary, within the new model of organisation, the global activities of large industrial groups are articulated according to a strategy, which may be defined of "localization" or of "local-but-global", which is based on the tight integration of subsidiaries localised in individual foreign countries or in various large continental regions in the perspective of a global strategy.

The various products are grouped in broad categories in order to identify a focus allowing to integrate among themselves different productions. Then, these different lines of products are allocated to the different foreign subsidiaries in a balanced way.

Each of these production units becomes responsible for a specific group of products at the global level or at least within a large transnational region, which may encompass various regions and countries. Therefore, each unit abroad has a much larger power than in the past and it is rather independent not only in the marketing and service to clients but also in the purchase of components, in the production and even in the development of new products.

This implies a shift from a vertically integrated structure, which was separated by the national borders, to a horizontally integrated structure operating at the transnational level.

The globalization of these different functions within each groups of products allows a greater efficiency, as it pools together resources, merges responsibilities and avoids duplications and conflicts and internal turf wars within the large multinational groups.

In fact, large firms should avoid those centralistic and bureaucratic organisations, which imply the risk of becoming too slow to manage the fast changes required by the market. A decentralised structure is more efficient, as each field of business is managed at the global level by only one subsidiary of the firm.

The decentralisation abroad of productions is convenient not only in productions with high labour content, but also in order to have access to a local specialised know-how and to suppliers, which have a higher quality level in a global perspective. Decentralisation helps to make the research and development programmes of new products more efficient and it aims to exploit new factors of efficiency trough production plants, which may serve overall continents rather than an individual country. Global headquarters only perform a supervision function of the management of the regional subsidiaries.

The increasing product differentiation and customisation are made possible by the standardization of the components and the mass production of these latter in plants which are localised in different countries and which serve a transnational market area.

However, since within these large transnational regions the preferences of consumers are rather homogeneous, the localisation within the individual countries and regions is not decided anymore in order to serve a specific captive market but according to the supply of production factors and the existence of organisational and technological capabilities, which are appropriate in order to exploit the emerging business opportunities.

The geographical contiguity or the direct presence within the individual countries is made necessary to increase the speed of reaction to these opportunities. Moreover, the capability to adapt to local tastes is becoming more important than in the past as the requirements of the customers are becoming more complex.

The development of a transnational structure makes strategic some requirements or constraints. In particular, it is necessary to develop a highly sophisticated logistic system and develop an information system integrated with those existing within each foreign subsidiary.

The presence of a foreign firm in each country should be stable and last for a period capable to insure a local image and to develop tight relationships with the respective local production system. For this reason foreign firms often acquire local firms, local trademarks and distribution structures.

The acquisition of foreign firms having a world known trademark is often finalised to a diversification of productions and to promote the diffusion of the firm at the world level into new sectors of activity.

In particular, large multinational firms have recognised the importance of having subsidiaries located in those highly specialised Italian industrial districts (for example: white appliances or sport shoes), which are world leaders in specific segments of production and which can represent an important source of external economies and have often decided to concentrate in these SMEs productions, which were previously scattered in the world. SMEs have benefited of a direct access to a large distribution organisation, to higher financial resources and to technological transfers and this has promoted an increase of employment in these firms.

The Internationalisation Strategy of SMEs. This new model of organisation, which may be described as a network, suggests that also SMEs may aim to perform a global role or at least be tightly integrated with other SMEs in foreign countries within a large transnational region. Thus, the increasing integration of the international and European industrial system and its structural transformation create new challenges but also new opportunities for SMEs (Cappellin, 1990, 1995b).

Internationalisation requires the capability by the firms to work in different environment and a greater decentralisation of functions and the creation of flexible alliances with foreign firms. A decentralised structure is more efficient than a concentrated one, as each field of business may be managed at the worldwide scale by an individual unit and this latter may be a small and medium size firm integrated in an international network with other SMEs.

In fact, the results from a series recent studies (Cappellin, 1995d) based on direct interviews of many large and medium size firms in regional economies highly characterised by the role of industrial SMEs, such as the regions of North Italy (three-Venetia regions and Lombardy), indicate that the trends observed in the spatial organisation by SMEs are highly similar to the network model increasingly adopted by large firms.

In particular, the internationalisation process of SMEs can be interpreted as the extension at an international framework of the same model of specialisation and cooperation with other firms, which since long time exists within a regional framework (Becattini, 1990; Cappellin, 1988).

A major characteristics of the internationalisation process of SMEs is the fact that it is a gradual "learning process" where the forms adopted by the individual firms vary continuously, trying to adapt pragmatically to the different environment of the various countries on the base of experience.

The internationalisation of SMEs through commercial agents or self-owned foreign commercial offices or subsidiaries is more important than that through foreign investment in production structures, although the development of international subcontracting agreements and the sale of licenses often represent a valid alternative to these productive investments.

Internationalisation requires a high decentralisation of operative functions of these firms and the creation of flexible alliances with foreign firms, which need to be monitored through a direct presence in the foreign country considered. In this perspective, the major obstacle to the internationalisation of SMEs is internal to the firms themselves and it is represented by the lack of qualified human resources or by the frequent existence of a too parochial culture, which is too different from that in the foreign countries considered.

Moreover, the lack within the SMEs of many qualified technical collaborators to the entrepreneur hinders that systematic effort, which would be required by the implementation of a medium term strategy.

The production capabilities of SMEs are largely incorporated in the qualification of the local labour force, in the local network of suppliers and in other factors characterising the local environment, which may hardly be reproduced abroad.

Therefore, SMEs often prefer to maintain production concentrated and are reluctant to decentralise it in less developed European regions and countries. This is especially true in the case of products, the demand of which is unstable and for which time and quality is the crucial factor of competition, as it is typical for SMEs operating in very specific segments of the international market.

On the contrary, SMEs usually prefer to invest in marketing structures especially within the most developed countries, aiming to integrate the value added which is created in the phases, which are downhill with respect to the manufacturing phase.

Only the firms, which maintain their specialisation in traditional productions, where the competitivity is mainly determined by costs, are interested in creating direct investments in foreign production plants in less developed countries. These firms often develop subcontracting agreement with firms of less developed countries, in the Far East, in East Europe or in North Africa countries, in order to dismiss production lines where costs have become too high.

3 The Microeconomic Characteristics of a Network

Various analysis of the networks of firms in various regions and at the international level indicate that the peculiar characteristics of a network with respect to other organisational forms are the followings (Cappellin, 1995b).

a) *A dynamic character*. Networks are characterised by a high flexibility with respect structural change or by an intrinsic dynamism or evolution of their specific organisational forms. The internal structure of a network and its relationships with outside evolve continuously according to the internal needs of its partners and the opportunities and threats created by the external environment at a faster pace than in the case of the organisational form: "large firm" and according to a more coherent pattern than the organisational form: "market". Therefore, networks insure lower "adjustment costs" to organisational change.

b) *Common aims, joint investments and leadership*. The internal organisation of a network is based more on the concept of leadership, of consensus and of a common identity, than on the concept of authority or of competition. A network is similar to a joint investment by the individual firms, which are willing to partially limit their respective autonomy in the perspective of the achievement of a common objective.

   Networks imply a production technology based on a division of labour among different partners. Thus they are characterised by a deverticalisation of production processes in different phases and the exploitation of "specialisation economies". They aim to exploit "economies of scope" and complementarity relationships in the common use of the respective resources of the individual partners. Networks require "idiosyncratic investments" which aim to reduce "transaction costs" between the partners and give stability to the relationships among them.

   Networks are based on a greater decision-making autonomy than large firms and aim to promote the entrepreneurial capabilities of each production unit. However, networks are also characterised in most of the cases by existence of a leader or a firm, which could insure a financial and strategic coordination among the partners. This requires some concentration of information flows and of control functions.

c) *An external openness and spatial diffusion*. A network model of organisation clearly implies a decentralised organisation in the territory. Thus spatial diffusion of plants may occur in various regions and countries or within a more limited territory as in the case of "industrial districts".

Thus networks imply a greater openness with respect to the outside than hierarchical organisations. They are also characterised by a flexible type of organisation, which is usually made by an "internal circle" of fewer firms to which other firms are related, which may be also considered as a part of the network but with a lower intensity or only for specific functions.

This openness of a network model of organisation makes it most appropriate in the diffusion of innovations and in the solution of the problems determined by "external diseconomies or economies" between various regions or between various local production systems.

d) *The existence of government structures and infrastructures*. A network is characterised not only by a set of partners and relationships or flows but also by a specific set of structures, which should manage in a rather formalised way the tight relationships among the firms. These "government structures" of a network may have an immaterial nature, such as in organisational routines, or a material nature such as in specialised services and intermediaries specialised in logistics, technology or finance, which work almost exclusively for the partners of the network considered.

Through a commercial or monetary relations, these services may insure the access by the various firms of a network to the same advantages, which usually pertain to the organisational form "large firm", due to existence of "economies of scale" or "network economies", while preserving to the firms an autonomy, which is very similar to that of a "market" organisation.

In a territorial perspective the impact of a network model of organisation is particularly important as it leads to:

* an increasing importance of physical infrastructures both traditional and very innovative,
* an increasing importance of business service activities, which may be considered as "soft" infrastructures,
* an increasing importance of urban centres, which perform the role of nodes or gateways in the relationships of the regions with the outside world,
* an increasing need for a change in the institutional procedures to be adopted within regional policies and of flexible mechanisms of coordination between the regions, the national governments and the European institutions, as also in the relationships among the various regions themselves.

4 Federalism and a New "Market Oriented" Regional Industrial Policy

The adoption of a network approach implies not only a change of the policy objectives and of the policy instruments, but also a change of the institutional framework in the decision and the implementation of regional policies.

4.1 The Aims and Instruments of a Modern Regional Industrial Policy

A modern industrial policy should first aim to be "market friendly". This implies to:

* use instruments of economic policy, which do not distort the market mechanism;
* correct the inefficiencies determined by case of market failures;
* correct through redistributing instruments the development disparities determined by the market mechanism.

A new "market oriented" regional policy should aim also to the following general objectives (Cappellin, 1995c):

* promote the creation of new internationally competitive production activities;
* determine the environmental and institutional preconditions to create market mechanisms where these are in fact not existent;
* promote a greater integration among the various actors by removing information barriers and transaction costs, which hinders cooperative behaviour.

In fact, the dynamism of productivity and the development of the capabilities to compete at the international scale is the fundamental criterion of the choice of a development strategy for regions tightly integrated in the European economy.

In particular, the competitivity of each individual firm depends on the competitivity of the territorial production system, in which it is embedded, on the qualification of the local labour force, on the relationships with its subcontractors, on the system of producer services and of local infrastructures and on the efficiency of the public institutions.

Thus, the goals of a modern, effective and rigorous regional industrial policy, different from the mere attraction of external investments by large firms, could be to promote (Cappellin and Tosi, 1993):

* *total quality* or to promote the creation of consumer oriented production activities and not of activities dependent on the public demand;

* *time competitivity* or to stimulate the flexibility of the local environment and to accelerate the pace of structural change;
* *internationalisation* or to promote a greater external openness of the local economy, of regional exports and of interregional technological transfers.

The policy instruments, which seem most appropriate to reach the above indicated objectives of regional policies, do not consist only on financial incentives to individual firms, but are more diversified.

In particular, public financial support to private investments rarely influence in a fundamental way the investment decisions of small and medium size firms and they certainly do not determine the medium-long term development of these firms. On the contrary, industrial policies should aim to foster the internal organisational and technological capabilities of the firms or the existence of favourable factors in their local environment. In fact, the propensity to invest by small and medium size firms depends first on their past profit accumulation and on the future profit expectations and only secondly on the cost of credit.

From defensive schemes of financial intervention in crisis ridden or so called "strategic" firms and sectors, which were traditionally managed by the national administrations, there is the need of a shift to horizontal policies capable to increase the international competitivity of the local environment within which the firms operate, such as policies in the field of services and infrastructures.

Therefore, the following broad set of regional policy measures may be identified:

a) *Measures on the internal capabilities of firms:*
- promotion of birth of new firms,
- vocational education,
- applied research,
- receptivity towards innovation.

b) *Measures on the relations or networks at the regional level*:
- promotion of subcontracting relationships,
- identification of new actors in the regional production system,
- promotion of initiatives by local actors,
- promotion of the creation of modern business services,
- promotion of the creation of modern logistic intermediaries,
- organisation of the territory and of infrastructure networks,
- promotion of new forms of public-private cooperation,
- reorganisation of public administration.

c) *Measures on the relations or networks at the interregional level:*
- promotion of exports,
- promotion of investments in marketing organisation abroad,

- promotion of production investment abroad,
- promotion of cooperation with foreign firms,
- promotion of a policy of regional marketing,
- promotion of interregional and transnational cooperations.

4.2 The Principle of Functional Subsidiarity and of Public-Private Cooperation

This shift in the strategies of regional policy leads to the need to identify a new institutional and organisational framework of the decision-making process.

The model of subcontracting and of networks between firms may indicate a new role of public institutions within a territorial production system. The partnership model, which regulates the relationships between the subcontractor and the main contractor or between the customer and the supplier, may be applied also to the relationships between public and private organisations (Cappellin, 1995c, 1995d).

The subcontracting relationships are characterised both by a conflict of interest, which allows to check the overall performance of the relationship, and by an increasing cooperation, which transforms both actors into partners interested to increase their joint competitivity with respect to external competitors.

These relationships of cooperation or conflict among firms producing different outputs seem to be more important factors, than competition among firms producing the same output, in the explanation of the industrial development of many European regions.

Thus, rather than to aim to extend the model of competition to the public sector, especially in those fields where privatisations of public services cannot allow an effective competition, a strategy of decentralisation of public policies from the State to other various collective institutions according to the principle of functional subsidiarity seems a better alternative.

In fact, decentralisation not only stimulates specialisation and flexibility, but also those conflict of interests or check and balances, which are the basis of a liberal and democratic system.

In particular, the legitimacy of national and of regional public intervention in the field of regional policy is challenged by the appearance of new public and private actors.

The principle of subsidiarity and of decision-making decentralisation should be implemented not only in the relationships between the State and the Regions, but also between the regional administration and the different public and private organisations, which exist within a specific region (Bianchi et al., 1995).

The crucial issue is to promote an original role of the different collective organisations in a process of self organisation of the regional society and economy. This strategy coincides with the shift from the hierarchy model of government to the network model, based on partnership.

Measures of regional policy should be promoted and implemented by all regional actors according to their respective aims and their values. Each actor has a different and complementary role, while no single actor, even when it is a public one, represents the privileged leader in the new regional policy. Different from technocratic and top-down economic planning, the new regional policy should be the result of a process of negotiation among the various actors and be made by the integration of the actual policies of the various actors defined according to a common strategy (Olson, 1992).

This evolution is especially needed in those fields of regional policy, where private firms, chambers of commerce, entrepreneurial associations and other collective organisations could cooperate with the national and regional administration such as in the technological transfers, the promotion of exports and tourism, the creation of infrastructures and in general in the promotion of economic development.

The wide diffusion of public intervention in many less developed regions, induced by traditional regional policies, has implied a slow down in the timing of decision making processes. In fact, the main obstacle to economic development in these regions is represented by the slow pace of public decision processes.

This contrasts with the fact that competitivity in the private sector is mainly determined by the shorter time to markets, by just in time in the coordination between different production units and by the development of modern logistic systems capable to reduce the time of the production cycles from some months to few weeks or even few days, to timely satisfy the demand in the various markets of a firm.

A modern regional policy requires that the functions of public institutions are neither the direct ownership and management of production activities and services nor the legal regulation through framework laws and the creation of "ad hoc" High Authorities. Public institutions may instead have the role of an "active interface" or of a "system integrator", capable to integrate the policies of the various regional actors, to stimulate their project design capabilities through the proposal of specific initiatives and to supply technical assistance in the implementation of the various projects.

Public intervention is needed when there are information asymmetries. In fact, the increasing decentralisation of the economy and the complexity of a modern industrial society create new needs for an integration function, which could be performed by public institutions as also by various collective organisations.

The working of modern economies and societies seems to be based on three elements, such as: the market mechanism, the hierarchical organisation of firms and various intermediary organisations and institutions, which define the "environment", which allows an economic development process (Turaine, 1995).

In particular, the role of public administration within a regional economic system and the relationships between the public and the private sector seem to have changed and to be tightly related with the evolution in the technological and organisational paradigms of firms, as it is suggested by tables 1 and 2.

Table 1 Three phases of development and organisational models

	'50ties	'70ties	'90ties
Organisational paradigms	fordism	individual entrepreneurship customer orientation	integrated systems just in time
Production units	large firms	small firms	networks
Organisational objectives	growth	value of differences	put order in the heterogenity
Performance criteria	production costs	quality of products	time of processes
Factors of competitivity	economies of scale	flexibility and in house product and process innovation	synergy, networking organisational innovation

Table 2 Three phases of development and institutional models

	'50ties	'70ties	'90ties
Institutional paradigms	centralism	regional autonomy self centred federalism	networks of contracts cooperative federalism
Principles of interdependence	hierarchical coordination	principle of subsidiarity	principle of cooperation
Role of Public actors	public ownership interventism	legislative regulation jurisdictional activity	integration collective entrepreneurship promote the actors

In fact, the model of formalised and comprehensive planning, based on macro-sectoral accounting, of dirigist or top-down policy-making and of public ownership of production firms corresponded to a phase where large firms had a "fordist" structure and mainly aimed to a quantitative growth and to exploit economies of scale.

The concentration of power at the central level implied the need of detailed prescriptions to guide the behaviour of decentralised branches of public administration. This has led to a bureaucratic organisation, where local branches of public administration were hindered to analyse the needs of the public and to take autonomous initiatives aiming to satisfy them.

On the contrary, a greater emphasis on local institutional autonomy, still temperated by the existence of national framework laws, seems to be coherent with a shift to a "laissez faire" ideology ("get the prices right" or "get the property rights right") where the paradigm was decentralisation, reliance on automatic market mechanisms, abolishment of subsidies, privatization and establishment of "high authorities" which should regulate the market competition. This model of policy-making corresponded to a new development phase, when the crucial factors of competitivity became the individual entrepreneurship capabilities, the customer orientation, the quality control, the accountability for results, the fiscal autonomy, the injection of competition into public services, the enhancement of differences, the adoption of process innovation. All these factors assume a paramount importance in those economic sectors and geographical areas where SMEs are particularly important.

Finally, a greater emphasis on institutional reforms ("get the institutions right") and on new organisational models, such as strategic alliances, public-private partnership, intergovernmental cooperation, public entrepreneurship and enhancement of the participation by local actors to the policy making process corresponds to the requirements of the actual economic and social system, where the system integration, the time to process and to market, the just in time, to put order in the heterogeneity, the enhancement of synergies and the adoption of product and organisational innovation are the crucial factors of regional competitivity. Thus, the role of government becomes that of catalysing new solutions through society, by integrating the complementary resources of the various actors.

Public institutions should promote the circulation of information, propose feasible plans of integration among the various private actors, when there are multiple suboptimal equilibria between the market forces. They should provide financial incentives to a cooperative behaviour of private actors and perform the function of monitoring and of control in the case of opportunistic or "free riding" behaviour.

While the second model may have been more appropriate for a phase of economic crisis, where industrial restructuring and cost saving was an imperative, it may not been adequate in a phase of economic recovery where the time perspective of actors returns to favour the medium and long term. In this different phase, the need to promote large programmes and projects in infrastructures and public services underline the need for a greater public-private partnership.

Therefore, according to a network paradigm, a new regional policy should be characterised by the explicit adoption of a "transactional" approach, where, differently from a "prescriptive" approach, the most important question seems to be not "what" has to be done, but rather "how" and with "which" partners.

A correct equilibrium between the free market and the role of the State requires that the public actors, such as the national and the regional administrations, define general directives of economic and territorial planning and will become the arbiters in the conflicts among the various regions or local areas.

On the contrary, private actors and local subregional authorities should have the responsibility of taking initiatives for the implementation of operative projects. A specific and rather autonomous function should be insured to each actor, both in the phase of policy design and in that of policy implementation, in the framework of integrated actions.

In this perspective, the approach of regional economic planning does not necessarily contrast with the approach of a regional "market friendly" economic policy. In fact, both these approaches have undergone an evolution. The approach of a "market friendly" regional economic policy now recognises that the crucial instruments of competitivity of the local production system are represented by the improvement of product and service quality and by the receptiveness to innovation, rather than policy measures aiming only to increase production efficiency and on cost advantages, such as industrial restructuring and financial or fiscal incentives to firms.

On the other hand, the regional planning approach has evolved from typically centralistic model based on the principle of authority to a model based on partnership and allowing a greater decentralisation of the decision making mechanism.

If these two alternatives of each individual approach are combined among themselves, as it is suggested in table 3, they indicate four different models of public governance of a regional economic system.

Table 3 Four models of regional governance: regional development strategies and regional institutional instruments

		Public Planning Modell based on	
		Principle of Authority	Principle of Partnership
Market Compet-itivity based on	Quality En-hancement	innovation centralisation ------------------------------- development agencies, master plans, project evaluation	innovation decentralisation ------------------------------- planning contracts, strategic planning, public-private cooperation
	Cost Minimi-zation	efficiency centralisation ------------------------------- public firms, administrative planning, welfare system	efficiency decentralisation ------------------------------- privatisation, outsourcing, authorities

The first case is that characterising a traditional Welfare State model, where the values of efficiency and centralisation lead to very traditional instruments, such as public firms, administrative planning, transfers to households and public services. A positive example of this model may be represented by the Nordic countries.

The second model is based on a technocratic approach, which recognises the need to introduce in a rather "dirigist" way some elements of innovation in the regional economic system. This model is characterised by the creation of development agencies, the elaboration of master plans and the use of project evaluation techniques. A clear example of this model is the French "amenagement" policy.

The third model, which has been particularly popular during the '80ties, is characterised by the objectives of decentralisation and efficiency and it has lead to privatisation and outsourcing to the private sector for the supply of various public services. This approach, according to which the market would be the most efficient allocation mechanism, has lead to the creation of Anti Trust Authorities, which would guarantee the competition, and other regulatory authorities, which define standards of service compatibles with consumer protection. A paradigmatic case of this model is that which has been mostly followed in the reform of the US public administration.

Finally, the fourth model is based on public-private partnership and it is characterised by the belief that decentralisation is highly related with organisational flexibility and receptiveness towards innovation. The coordination among the different and autonomous regional public and private actors is promoted through procedures of negotiation, the elaboration of strategic plans, the creation of planning contracts and joint ventures, consortia or "mixed companies". This model, which may be defined as that of a "market oriented" regional policy, seems to correspond to the characteristics of the German federal system and of the public governance approach adopted in many local production systems or industrial districts in Italy.

In this perspective the entire system of procedures of interaction among the various public organisations should be redesigned according to a systemic logic. It is clear that a "lean organisation" or a "learning organisation" requires an effective decentralisation of decision making. Administrative and fiscal decentralisation or federalism represents the instrument to "reinvent government" (Osborne and Gaebler, 1992) or to define an organisation of the public sector, which may be lean, effective and efficient.

In particular, planning both in private and in public organisations has still a crucial role (Mintzberg, 1994). In fact, is not the competition on the market, which could bring spontaneous adaptation through an implicit type of coordination, but there is the need for an explicit, conscious, deliberate, purposeful anddeeply intentional kind of coordination.

However, a modern regional policy cannot be based on plans enforced by law and through a hierarchical coordination by public institutions.

The new approach to regional policy should be based on the principle of partnership rather than on that of public authority and on the capability by the different private and public organisations to communicate one with the others and to promote common projects, which would allow them to jointly achieve also their individual aims.

The crucial issue is the management of complexity and of flexibility and this requires that public authorities identify general aims or very broad strategies of development, rather than many detailed objectives, in order to promote the consensus and the synergies of the many different regional actors.
National and regional public administration cannot aim to an hierarchical coordination between the policies of the various private and public actors and of the municipal authorities, but they should rather define flexible mechanisms of governance of the relationships between the various public and private organisations (Kreukel and Salet, 1993).

The concept of planning assumes a different form in a federal or decentralised system. Planning has to be understood as a sort of "incomplete contract" or as a joint investment decision, where various reciprocally recognized actors decide to pool some of their own resources for the achievement of common aims, while defining procedures of flexible coordination and creating the institutional structures within which the decision-making will take place. Thus planning becomes a process of "strategic governance" of contractual and institutional relations.

5 Federalism and the Network Paradigm

The creation of the regional administrations in Italy as in many other European countries during the '70ties was mainly inspired by a technocratic approach and the new regional administrations have received mainly competences in the field of economic and territorial planning.

This approach has led to the adoption of centralistic and top-down procedures in regional planning, since the political and fiscal power has been firmly stayed concentrated in national institutions.

On the contrary, regionalism in the '80ties and '90ties seems mainly characterised by political, historical, cultural and social factors and by the ambition of the individual local communities to exercise an autonomous role at the national and international level.

Therefore, regionalism is incompatible with a dirigist and technocratic approach and it seems complementary with the effort to mobilise the various actors of a modern and pluralistic economy.

A federalist reform of the relationships between the State and the Regions has first a political dimension, rather than a constitutional, fiscal and economic dimension.

In the particular case of Italy, a federalist reform seems to imply to:
a) enhance the regional identity and the national and international role of each region;
b) redefine the role of the public sector in the economy;
c) modify the characteristics of the actual regional policies of the State and the European Union.

5.1 The Reasons for Decentralisation

The analysis of federalism has been elaborated by orthodox neoclassical theory in terms of consumer preferences and of technological/production costs, focusing on the search of an optimal dimension of government units (Eser, 1995; Thoeni, 1986).

According to traditional consumer demand theory, decentralisation and federalism have been justified by the aim to increase the freedom of choice of the citizen/consumer on the level and typology of the public services. In fact, decentralisation allows a greater customization of public services according to the different needs of the various local production systems. This allows a greater level of political consensus on the policies to be adopted within more limited and homogeneous constituencies. These arguments justify a greater fiscal autonomy of local governments.

According to traditional firm production theory, decentralisation or centralisation have been justified according to the characteristics of the economies of scale and of threshold effects in the production of various public services. Similarly, the existence of external economies or diseconomies, the existence of spillover effects and of ruinous or wasteful competition suggest the need for coordination or of greater administrative units.

A model of "dual" or "competitive" federalism implies a clear-cut separation of competences founded on constitutional norms and it is founded on a clear division of "property rights" between the various institutional levels, as suggested by neoclassical theory . However, this seems a too narrow conception of intergovernmental relations (Haeberle, 1995).

Moreover, traditional neoclassical theory considers the case of the limits of local government and it has emphasized the factors explaining the need for an enlargement of local administrative units or for a greater centralisation.

However, the factors leading to a decentralisation are not only just the opposite of those leading to the incorporation of smaller administrative units into a larger one. In fact, modern industrial economics and theory of endogenous development allow to identify various arguments for decentralisation, which pertain to different methodological approaches and to a different body of economic research (Rallet and Torre, 1993; Cappellin, 1988, 1993b, 1995a). Clearly, the network paradigm emphasizes the need for greater decentralisation.

In particular, the decentralisation of a composite set of complementary functions to local governments could allow to exploit economies of scope and synergies.

A greater closeness to the local production system allows greater flexibility, shorter time to response and timely changes of policy priorities.

Decentralisation allows a greater access to local information networks and greater capability to identify strengths and weaknesses of the local economy. Thus, it allows a greater efficiency in collecting and in distributing information.

Decentralisation facilitates the sharing of common values and development strategies and allows to mobilise the entrepreneurial resources and forms of public-private cooperation in the production of particular public services (Putnam, 1993; Leonardi, 1993).

In particular, there are various reasons why industrial policy should be designed and implemented at the regional level (Cappellin and Tosi, 1993; Cappellin, 1993a, 1995a).

First, the process of deverticalisation of production processes determines a decrease of the ratio of value added/sales. Thus, the competitivity of firms in terms of unit costs is determined mainly by the capability to control the prices of intermediate goods and services, many of which are purchased from subcontractors and other suppliers at the local level and may then be influenced by appropriate regional policies.

Moreover, the complexity of the new model of organisation based on networks of firms and the development of modern logistic activities requires that industrial policies are adapted to the very specific characteristics of each production and technological system and are thus defined and implemented at the regional level.

Finally, an industrial policy aiming to promote innovation and the dynamic competitivity of the firms should be managed at the regional level, in order to better identify and remove the specific factors that hinder adoption of innovation especially in the case of SMEs (Cappellin 1993a, 1995b; Cappellin and Tosi, 1993).

Thus, the competitivity of each individual firm depends on the competitivity of the local production system in which it is embedded or on the qualification of the local labour force, the relationships with the system of local suppliers and subcontractors and the efficiency of the local public institutions.

These arguments may be summarised by stating that decentralisation allows:

a) *lower information costs*, as only strategic and not routine information has to be transferred to the centre;
b) *lower transaction or coordination costs*, as through the empowerment of individual local administrative units with an integrated set of various functions it is possible to help an integrated approach, to promote synergies and "networking" between the public institutions and between these latter and the private actors at the regional level and to avoid those conflicts usually existing when different functions or policy areas are attributed to separate departments of the central administration;
c) *greater receptiveness to change*, as decentralisation, greater autonomy and responsibility reduce bureaucratic behaviour, stimulate a result oriented mentality, allow experimentation and stimulate competition and the emulation of benchmarks also between public institutions.

5.2 An Organisational and Institutional Perspective to Federalism

The economics of federalism is different from that of the optimal size of local governments. The neoclassical theory of local government emphasizes the problem of the separation of competences between different vertical levels of administration, while the economics of federalism should examine the problem of the intergovernmental relations between different vertical levels of government.

Federalism can be studied with the analytical instruments of the modern organisational economics or the new institutional economics, focusing on the mechanism of contracting or on the analysis of different governance structures of intergovernmental relationships and their different efficiency.

In fact, institutions matter or are important and are persistently neglected in the planning of development (Williamson, 1994; North, 1990). "Modern economic growth results from the development of institutions that permit an economy to realize the gains from specialisation and division of labour associated with sophisticated technology that has developed in the Western world in the last several centuries" (North, 1987).

We may say that federalism and the need for a greater decentralisation of institutional structures correspond to a paradigmatic change of the structure of society, of economy and of production technologies in advanced industrial systems. Federalism insures greater decision-making decentralisation and it represents the institutional form, which is most appropriate for a socioeconomic system, which has become more articulated, more culturally developed and more technologically complex.

In particular, according to transaction costs theory (Williamson, 1981 and 1975) the purposes of federalism as that of any economic organisation may be defined as those to economize on bounded rationality and to mitigate opportunism in the relationships between different administrative units.

The regional decentralisation of a national public administration is similar to a "buy" (opposed to the "make") decision by a firm or to the externalisation of some functions, which were previouslyproduced internally. In fact, decentralisation implies a tighter relationship between the administrative unit delivering the service and its users in the various individual areas and this is preferable in the case of services implying:

- high frequency of transactions with the local users, such as with routine services;
- high complexity or need for flexibility to specific local demands;

- some degree of idiosyncratic investment or of asset specificity, such as in economic development policies based on public-private cooperation, which requires of credible commitment by both parties.

Thus, it is possible to apply to federalism the analytical tools, which have been devised in the analysis of contractual relationships, such as: opportunism, credible commitments, reputation, confidence, trust, consensus, sense of belonging, identity, strategy and other related concepts.

Federalism may be considered similar to an extension of the market model to intergovernmental relationships, whether it interpreted as an institutional system based on clearly defined property rights and on competition of the different regional constituencies.

Otherwise, federalism may be considered similar to the establishment of a network organisation in the relationships between the State and the Regions. In particular, a federal system may be considered similar to a "diversified organisation". These latter are organisations split into quasi autonomous divisions to serve a diversity of markets, with "headquarters" relying on financial control systems to standardize the outputs of the divisions (Mintzberg, 1994), as it typical of modern multinational corporations.

In fact, similarly to the matrix organisation of multinational firms, a local public agency responsible in a specific sectoral field may be obliged to report both to the national ministry having a coordination function in the same field over the national territory, as also to the regional government, which has the authority for the intersectoral coordination of the various sectoral agencies within the individual regional territory considered.

In particular, the network paradigm of organisation may complement the indications of transaction cost economics, as it explains various characteristics of a federal institutional model.

Transaction cost economics focus on the analysis of bilateral relations between two production or administrative units, while the network approach allows to analyse the case of multilateral relationships.

As it happens with the network of subcontractors of a firm, also the relationships between the local authorities, the Regions and the national administration should not be interpreted in a bilateral perspective, while the role of each actor should be understood in the framework of the overall network of intergovernmental relationships. Thus, transaction cost economics focuses on bilateral bargaining, while the network approach focuses on the concept of leadership and consensus building within a larger and semi-hierarchical system of various constituencies.

Moreover, according to the network approach, negotiations between different units do not represent only a transaction cost or a barrier for a smooth working of the economic or institutional system considered. On the contrary, they represent a crucial and positive search activity of a solution, which may be more rational and protect the parties considered from the other's opportunistic behaviour. Negotiations do not represent an obstacle, but rather, as a "catalyser", they increase the connectivity among the partners of a network and decrease the "adjustment costs" during the innovation adoption process and shift to more effi cient allocations of resources. Thus, the choice of an efficient governance structure is not that which allows to economize on negotiation efforts or on transaction costs, but rather that which is capable to secure appropriate efforts in the negotiation process, which represents a form of "idiosyncratic investment" and allows to maximize the results of the relationships between the various institutions.

In fact, an increasing information complexity and the need for timely decisions require a decentralisation of responsibilities. On the other hand, a decentralisation of responsibilities requires a higher access to information, a progress in communication and greater management capabilities of the various economic and institutional actors.

Federalism as other network organisations is characterised by high flexibility. Thus, federalism represents a learning process rather than a model and it is not a set of laws applicable to all intergovernmental relations, but a mechanism or a highly adjustable framework for institutional arrangements in the State-Regions relationships.

These institutional arrangements govern the ways through which the national and regional units can cooperate and/or compete. They provide a mechanism that can determine a change in the laws and may redefine the "entitlements" (Dahrendorf, 1988) or the "property rights" of the various constituencies.

This flexibility of a federal system explains thedifficulty in defining in legal terms the implications of the principle of subsidiarity. That should not be understood as a limit only of this principle, as these characteristics are similar also to other key principles, which may be applied to a whole range of contractual transactions and which, for their generality and competing character, should not be stated as legal rules at all (Williamson, 1994).

The relationships between the various level of government vary in time or adapt continuously to new internal and external changes. The management of intergovernmental relations seems more an art than the implementation of scientific analysis.

As it occurs in other network organisations, a federalist structure represents a dynamic process of transformation, which may proceed gradually from the existing schemes of administrative decentralisation, rather than be the result of a "revolutionary" change of culture, leading to centralistic counteractions or to secession, which represents a form of micro-centralism.

In the case of Italy, instead to discuss on the different models to be adopted from other federalist countries it seems more opportune do adapt since now the existing Italian regional system, on the base of the more than twentyfive years of experience of the regions with "special" and "ordinary" competences.

5.3 The Limits of the Principle of Subsidiarity

An institutional and dynamic approach to federalism underlines a limit of the well know subsidiarity principle. This principle, which characterises a federal system, implies that competences and responsibilities should be attributed to that institutional level, which is most close to the problem considered and is suitable for its solution, corresponds to a reasonable approach aiming at the decentralization in decision making and the accountability of individual actors or institutions (Haeberle, 1995a; Ridola, 1994).

The principle of subsidiarity represents the base for a "competitive federalism" model, which implies a clear separation of the competences attributed to each level of government. Thus, the subsidiarity principle seems to be still inspired to a vertical or hierarchical approach (Cappellin, 1993b, 1995a) and it contrasts with the network approach, which aims to a more flexible organisation of intergovernmental relationships, based on the principle of cooperation or partnership.

In this perspective, the subsidiarity principle may be complemented by the development of operative procedures of vertical and horizontal cooperation. In particular, the paradigm of networks allows a shift from a federalism model of autonomy and division to the model of "cooperative federalism", which emphasizes the concept of synergy.

In particular, new approaches to regional policy should integrate the subsidiarity principle, according to which the management of policy measures should be devolved to the lowest level possible, with the networking approach, which allows to enhance the powers or the duties of setting guidelines by national or E.U. institutions and the powers or the rights of initiative of the regional institutions and the various actors at the regional level.

Thus the national and EU institutions could define broad strategic lines and identify objectives capable to create a consensus between the various institutions and organisations involved, while the choice of the means to be adopted in achieving those goals could depend upon local institutions.

Federalism and decentralisation are needed also to qualify the functions of the central government, which would gain in terms of flexibility and could exploit economies of specialisation, by focusing its action on the fields of really national relevance.

On the other hand, an efficient and "lean" central administration may be a crucial factor, which could strengthen regional autonomy, as the development of decentralised decision-making is facilitated by the existence of a central actor, which could perform the role of a "system integrator" and insures a vertical and horizontal flow of information to the local administrations.

5.4 Regional and National Identities and Interregional Solidarity

National identity or a common sense of national belonging represents a sort of "public good" which allows to decrease the transaction costs of the intergovernmental relationships between the regions of the same country and between these latter and the national State.

Only the trust of the regional government in the national government may induce the various regions to partially renounce to their freedom or to their entitlements, as it is required by the logic of cooperation and integration.

Similarly to a partnership, the individual regional institutions may accept some limitations of their respective freedom, whether they trust that the complementarity of their respective resources will allow them to collect the benefits of a synergy or of "network economies".

National unity allows to stabilise the relationships between the various regions on the base of the individual specialisation and of the access to complementary resources of the other regions and it may facilitate each region in the aim to promote an endogenous development process.

On the contrary, the complete autonomy of a region becoming an independent state is similar to the establishment of an absolute property right. This may lead to isolation and competitive or conflictive solutions. A preferable alternative may be the federalist solution, which is similar to a weaker form of property right implying mutual interdependence and thus favouring cooperative relationships.

Federalism is not only an institutional framework, which allows a continuous or incremental process of negotiation through which it is possible to settle and change the individual rights, which regulate the interaction between the various regions and the central government. Federalism is also based on the acknowledgment by each government level of a common set of shared values and of a common perspective or plan of development.

National unity in a federal state implies the existence of ideological attitudes, which are based on a common perception of reality and a common view of fairness and justice of institutional arrangements. This allows the existence of credible commitments between the various regions and limits "free riding" or opportunistic behaviour. In particular, the sense of national belonging and interregional solidarity justify the existence of interregional transfers, as in the model of "unitary" or "trust" federalism (Haeberle, 1995a).

In fact, federalism, similarly to other forms of decentralisation such as the political democracy and the market mechanism, is a governance structure, which is based on the acknowledgement of the ethical value of the freedom and the rights of the individual regions.

Federalism is based on profound political assumptions, such as the fact it represents a guarantee of political and personal freedom, an instrument to defend the rights of all sorts of minorities, a guarantee of separation of powers and a mechanism of check and balance, which is typical of a legal state. Federalism is justified by the will to guarantee cultural pluralism and a strong autonomy in cultural policies is a fundamental factor of federalism, such as in the field of education, research, universities and arts and musea or in the field of media (Putnam, 1993; Haeberle, 1995b).

Thus, it seems fairly obvious to identify the difference between a region and a nation in the greater political powers of the nations (Krugman, 1991), as the powers and the financial dimension of the budget of regional governments widely differ and they may even be really large with respect to that of the respective nation. The crucial difference seems instead that international relations are regulated by short term agreements, which are similar to "complete contracts" mainly based on respective interest and on "ad hoc" common objectives. On the contrary, interregional relations are based on the reciprocal acknowledgement of a common identity or sense of belonging, which makes the interregional agreements similar to long term "uncompleted contracts" and leads the individual regions to accept the power of common national institutions.

A crucial factor at the foundations of federalism is the legitimacy of the various government levels, as each of them must be considered a credible representative of its respective constituency and simultaneously be considered a credible partner by the other government units, which participate to the same collective decision making process. In particular, the national government should be legitimatized by the consensus of regional governments.

The consensus on the national unity is based on the common awareness and thrust on the principles of justice or of solidarity, which may be considered as a sort of reciprocal interregional insurance against asymmetric shocks, which may strike an individual "crisis" region. Thus, the national budget does the function of a "shock absorber".

In particular, national regional policies or interregional financial transfers represent a fundamental element of a federal system, as it is suggested also by the increasing role of regional policies in the European institutions (Cappellin and Molle, 1988; Cappellin, 1993a). In fact, national regional policies make a federal state different from a system of completely independent national states.

However, a federal perspective implies some constraints in the forms of interregional solidarity.

First , the principle of decentralisation implies that of responsibility and accountability. Thus, a sound behaviour of the individual regional government should be induced by the existence of positive incentives and of sanctions. In this perspective, decentralisation requires either "fiscal federalism" or fiscal autonomy. In fact, the existence of clear financial rules and of a financial discipline in the intergovernmental relationships contributes to avoid opportunistic or "free riding" behaviour by some regions.

Secondly, according to the principle of subsidiarity, interregional transfers towards the "crisis" regions should aim to enhance the endogenous capabilities to face an eventual critical situation and they should be mainly oriented to promote internal investments, rather than aim to sustain the personal incomes and consumption levels.

Thus, a further similarity between federalism and a network organisation is represented by the fact that federalism, such as a network, may be interpreted as a joint investment between various regions, which decide to pool some of their resources in order to gradually realize a common aim.

We may conclude that federalism or decentralisation implies three complementary factors, which respectively have a subjective, an objective and a collective character. First, it implies that each constituency shall be characterised by a set of provisions, human and natural resources, competences or production capabilities. Secondly, federalism represents an institutional framework, which allows to clearly define and enforce the entitlements of each individual government levelor the objectively recognised "property rights", which are at the base of the game of intergovernmental relationships. Finally, federalism implies a collective dimension or a common sense on belonging, which induces reciprocal thrust and solidarity.

6 Conclusions

The recent changes in the technological and organisational structure of firms, the increasing consensus on a "market oriented" role of public institutions and the request of a greater decentralisation in the State-Regions relationships suggest new opportunities and not only new constraints for regional policies.

First, the internationalisation process of firms requires a high decentralisation of the operative functions and the creation of flexible alliances with foreign firms. Large industrial groups are becoming articulated according to a strategy, which may be defined of "localization" or of "local-but-global".

Internationalisation requires the capability by the firms to work in different environment and a greater decentralisation of functions and the creation of flexible alliances with foreign firms.

Secondly, there is a tight relationship between federalism and the modern forms of organisation in the individual firms and the development of cooperative relationships among the various local actors within a local production system.

A modern regional policy suggests that the role of public institutions is neither the direct ownership and management of production activities, nor the simple regulation of free competition, but rather the aim of representing a "system integrator" and of taking the initiative of strategic projects requiring the cooperation of the various regional actors.

The paradigm of networks represents a useful methodological instrument to interpret the relationships between the firms and those between the government and the private organisations. According to a network paradigm, a new regional policy should be characterised by the explicit adoption of a "transactional" approach, which may substitute the traditional "prescriptive" approach.

Thirdly, there is the need to define a new institutional framework of the relationships between the State and the E.U. institutions, on the one hand, and Regional and Local Authorities, on the other hand.

A networking approach allows to enhance both the powers or the duties of setting guidelines by national or E.U. institutions and the powers or the rights of initiative of the regional institutions and the various actors at the regional level.

In particular, it suggests why administrative and fiscal decentralisation or federalism may represent a more effective approach than the adoption of the competitive market model and it may be crucial in a strategy aiming to "reinvent government" or to define a new organisation of the public sector, which may be lean, effective and efficient.

References

Becattini, G., The Marshallian industrial district as a socio-economic notion, in F. Pyke, G. Becattini and W. Sengenberger (eds.), Industrial districts and inter-firm co-operation in Italy, Geneva: International Labour Organisation - International Institute for labour studies, 1990.

Bianchi, P.,Le politiche industriali dell'Unione Europea, Bologna: Il Mulino, 1995.

Bianchi, G., R.J. Groete and S. Pieracci, Dalla coesione economica alla coesione istituzionale, Sussidiarietà funzionale e reti socio-istituzionali nelle politiche regionali, in G. Gorla and O. Vito Colonna (eds.), Regioni e sviluppo: modelli, politiche e riforme, Milano: Franco Angeli, 1995.

Cappellin, R., Transaction costs and urban agglomeration, Revue d'Economie Regionale et Urbaine, n. 2, 1988.

Cappellin, R. The European Internal Market and the Internationalisation of Small and Medium Size Enterprises, Built Environment, 16, 1, 1990.

Cappellin, R., Patterns and policies of regional economic development and cohesion among the regions of the European Community, in R. Leonardi (ed.), The state of economic and social cohesion in the Community prior to the creation of the Single market: the view for the bottom-up, London: The European Institute, The London School of Economics, 1993a.

Cappellin, R., Interregional cooperation in Europe: an introduction, in R. and P. Batey, (eds.), Regional Networks, Border Regions and European Integration, London: Pion, 1993b.

Cappellin, R., Regional development, federalism and interregional co-operation, in Eskelinen, H. and F. Snickars (eds.), Competitive European Peripheries, Munich: Springer-Verlag, 1995a.

Cappellin, R., Subsidiarity as a new economic strategy, in M. Steiner (ed.), Regionale Innovation: durch Technologiepolitik zu neuen Structuren, Graz: Leykam, 1995b.

Cappellin, R., Una politica regionale nazionale "orientata al mercato" tra i nuovi modelli organizzativi e il federalismo, in Gorla G. e O. Vito Colonna (ed.), Regioni e Sviluppo: modelli, politiche e riforme, Milano: Franco Angeli, 1995c.

Cappellin, R. Regional embeddedness and international integration: the case of SMEs in North Italy, paper presented at the National Congress of the Associaçâo Portuguesa para o Desenvolvimento Regional, Porto, 27-29 April 1995d.

Cappellin, R. and Molle, W., The coordination problem in theory and policy, in Molle, W., and Cappellin, R. (eds.), Regional impact of Community policies in Europe, Aldershot: Avebury-Gower, 1988.

Cappellin, R. and A. Tosi, (eds.), Politiche innovative nel Mezzogiorno e parchi tecnologici, Milano: Franco Angeli, 1993.

Coase, R.H., The institutional structure of production, American Economic Review, n. 82, 1992.

Dahrendorf, R., The modern social conflict, An essay on the politics of liberty, New York: Weidenfeld & Nicholson, 1988.

The Economist, A survey of multinationals, June 24th 1995 and March 27 1993.

Eser, T.W., A normative economic theory of subsidiarity and the evaluation of the regional policy in the West Middlands, paper presented at the 35th European Congress of the R.S.A., Odense, August 21-25, 1995.

Kreukels, A.M.J, and W.G.M. Salet (eds.),Debating institutions and cities, The Hague: Netherlands Scientific Council for Government Policy, 1992.

Krugman, P., Geography and trade. Cambridge, Mass.: MIT Press, 1991.

Haeberle, P., Das Prinzip der Subsidiarität aus der Sicht der vergleichenden Verfassungslehre, Universität Bayreuth, unpublished manuscript, 1995a.

Haeberle, P., Unità politica e pluralismo culturale in Germania, in A. D'Atena (ed.), Federalismo e regionalismo in Europa, Milano: Giuffrè, 1995b.

Leonardi, R., Economic and social cohesion and the Community's Institutional architecture: the role of transregional networks, in R. Leonardi (ed.), The state of economic and social cohesion in the Community prior to the creation of the Single market: the view for the bottom-up, London: The European Institute, The London School of Economics, 1993.

Mintzberg, H., The rise and fall of strategic planning, Hertfordshire, Prentice Hall International, 1994.

North, D., Institutions, transaction costs and economic growth, Economic Inquiry, July 1987.

North, D. C., Institutions, institutional change and economic performance, Cambridge: Cambridge University Press, 1990.

Ohmae, K., The borderless world, New York: Harper & Row, 1990.

Olson, M., The rise and decline of nations, New Haven: Yale University Press, 1992.

Osborne, D. and T. Gaebler, Reinventing goverment, Reading, Mass.: Addison-Wesley, 1992.

Putnam, R.D. (with R. Leonardi and R.Y. Nanetti), Making democracy work, Civic tradition in modern Italy, Princeton: Princeton University Press, 1993.

Rallet, A. and A. Torre, Regional and industrial economics at the crossroads, paper presented at the 33rd European Congress of the Regional Science Association, Moscow, 24-27 August 1993.

Ridola, P., Das Prinzip der Subsidiarität im Regionalstaat, Die Beispiele Spaniens und Italiens, Eine Kritische Würdigung, in Nomos (ed.), Susidiarität, Nomos Verlagsgesellschaft, Baden-Baden, 1994.

Scott, A.J. and M. Storper, Regional development reconsidered, The Lewis Center for Regional Policy Studies, University of California at Los Angeles, Working Paper n. 1, 1990.

Thoeni, E., Politökonomische Theorie des Federalismus, Baden-Baden: Nomos Verlagsgesellschaft, 1986.

Touraine, A., L'impresa e lo Stato nella società post-industriale, in P. Bassetti (ed.), Impresa & Stato, Bologna: Il Mulino, 1995.

Williamson, O.E., Market and hierarchies: analysis and antitrust implications, New York: The Free Press, 1975.

Williamson, O. E., The modern corporation: origin, evolution, attributes, Journal of Economic Literature, 19, 1981.

Williamson, O. E., Institutions and economic organisation: the governance perspective, paper presented at the Annual Bank Conference on Development Economics, World Bank, April 28-29, 1994.

Inequalities and Mobility in the Danish Welfare State

Nils Groes
Amternes og Kommunernes Forskningsinstitut (AKF)
Nyropsgade 37
DK - 1602 Copenhagen

1 Summary

The paper deals with the economic development in different parts of Denmark – so far a welfare state with ever falling inequalities.

In spite of the decline in employment in agriculture and fisheries, the living standard has not increased less in the peripheral areas than in the rest of Denmark. On the average, rural districts have witnessed greater increase in employment and disposable income per capita than Denmark as such.

At least *three* explanations of the growth in income and employment in many rural municipalities can be found:

First, the *growth in social transfers* has been considerably larger than the growth in factor income during the last two decades. And especially rural municipalities (where factor incomes are low) have gained from this development.

Second, *taxes* have also risen more than factor incomes. Therefore, in municipalities where taxes are low, they have only taken a smaller piece of the income growth.

Third, *commuting* out of the rural districts has continued to rise. The number of local jobs has been stagnant, but more employees in the rural districts commute to other municipalities. Actually, the number of employed living in rural areas has risen. Short distances and a good road system in Denmark explain part of it.

Nevertheless, the industrial development is not taking place in rural areas and especially not in the peripheral municipalities.

Regression analyses show, however, no evidence of economies of scale (factor income and number of jobs growing most in large cities). There is no significant spatial concentration, when due regard is taken to the decline of employment in agriculture, the difficult position of the peripheral areas and the fact that high income areas tend to attract more inhabitants.

But we have a spatial economic paradox: while growth in employment in West Denmark (Jutland) has been substantially higher than in East Denmark, then the main part of immigrants and refugees has rushed to the East: to the metropolitan region where growth has been markedly smaller and unemployment higher.

We do have explanations: It is a question of our welfare policy, of the incentives to work, of housing policy and of cultural patterns. And the paradox just illustrates the general lack of mobility on the Danish labour market.

2 Small Differences

Driving through one of the smaller Danish islands it is difficult to believe that you are now in one of the poorest regions of the country – a peripheral area with low incomes, high unemployment and migration. Everything looks so nice and well-kept, it is difficult to catch sight of abandoned houses not to mention dead villages like the ones you see elsewhere in Europe. And the population is hardly distinguishable from other Danes as to height, breadth, colour and clothes.

The truth is in fact that although compared to the rest of the country the income is low in a peripheral area like West Lolland there are no big differences. In 1992, the disposable income (after tax) per capita was only 9% below national average. But the factor income was 24% below national average. The explanation is partly that taxes are about 20% lower than in the rest of the country, partly that pensions, unemployment benefits, cash benefits and other public transfers are almost 30% above national average. The social transfers are larger than the direct taxes on West Lolland, while the same taxes in Denmark as a whole are 50% higher than the transfers. In the well-to-do municipalities the direct taxes are more than three times as high as the transfers.

So when the living standard in peripheral areas does not seem discouraging low it is mainly due to lower taxes and bigger transfers – paid by the rest of the country. This is a natural consequence of the Danish welfare policy.

3 The Development

In Denmark, during the last decade the disposable incomes have had a real growth of almost 2% per year.

Figure 1 The growth of disposable real income per capita (1984-1992), all municipalities

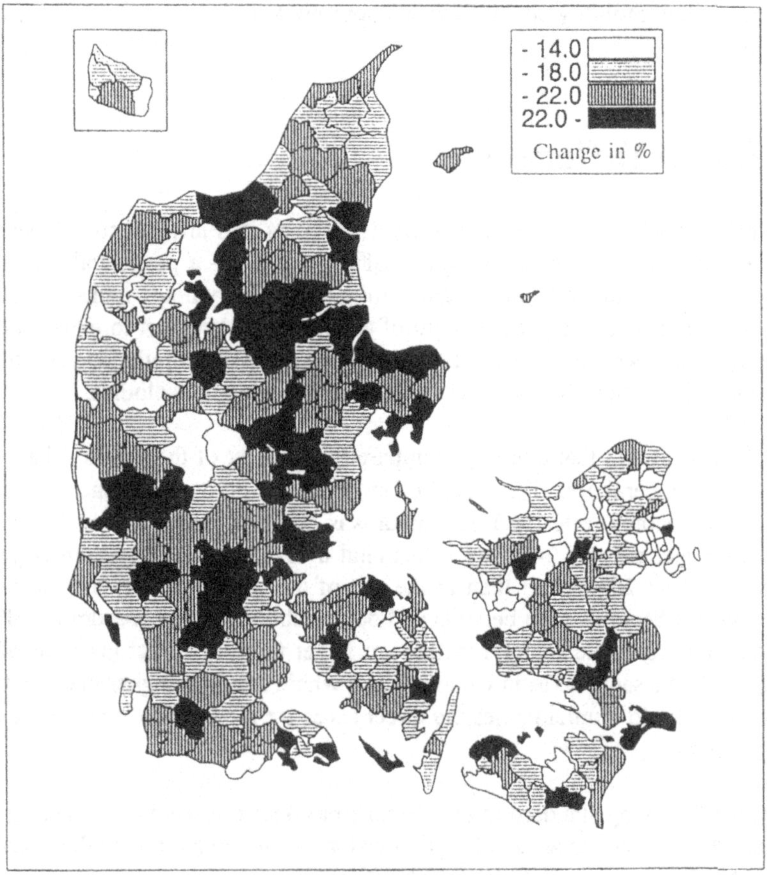

Source: AKF's Data Bank.

But during the same period social transfers have risen more than twice as fast in real terms. Taxes have also increased faster than income. The growth of these items has influenced the regional income distribution.

As shown by Rowntree (1995) Denmark actually belongs to the very few European countries with falling inequality in the eighties. The question is, however, whether this development will continue and if not, what the spatial economic consequences will be. But the future development is not the issue of this paper.

Figure 2 Population development from 1984-1992

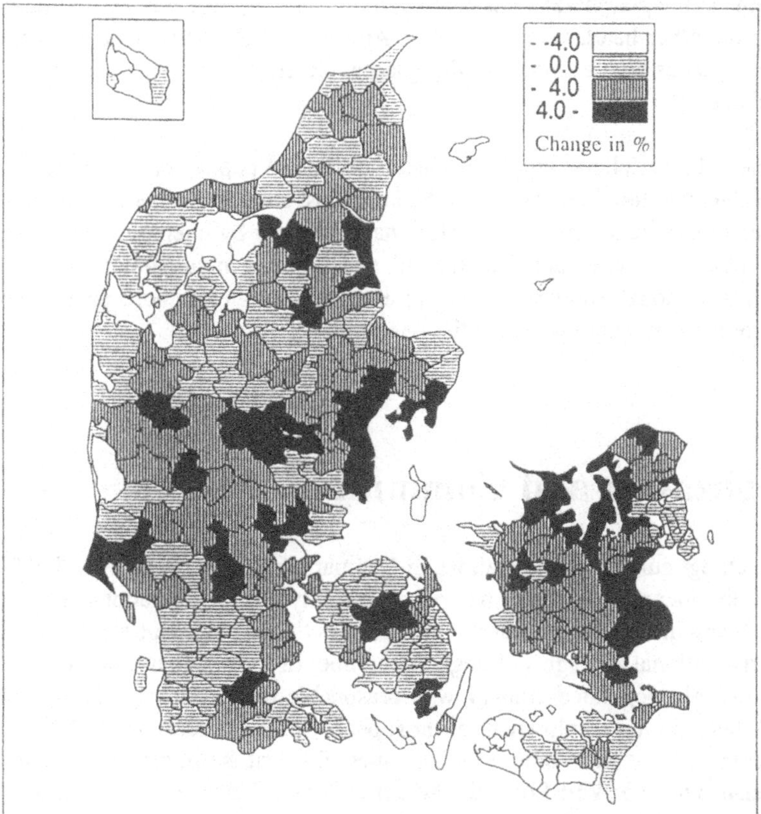

Source: National Statistical Office of Denmark.

From 1984 to 1992 the income per capita in the peripheral areas has grown just as much as in the rest of the country. However, the number of inhabitants has decreased a little. You could always discuss the definition of a peripheral area, and how many municipalities you have to include. Bjarne Madsen (1994) defines vulnerable municipalities as those situated a long way from dynamic economic centres. He arrives at municipalities including a total of 400,000 Danish inhabitants. During 1984-1992 these municipalities have had a growth in disposable income per capita of a little more than the national average, cf Appendix 3. Apparently, this is not due to social transfers having grown more in the vulnerable areas than in the rest of Denmark. But in these areas the transfers play a more important role, partly because numerically they are higher than in the rest of the country, partly because the factor income is smaller. Actually, the growth in transfer incomes amounts to more than half the total rise of income in the peripheral areas compared to only 1/4 in the well-to-do municipalities. On the other hand, taxes are lower. And even if the percentage growth of direct taxes is just as high as in most other places, the growth in DKK is only 3/4 of the national average.

As mentioned, the peripheral areas have had a decline in population and local jobs. The most vulnerable municipalities have had the most sincere decline and the problems for this group are accumulating. More have been hit by unemployment, especially long- term unemployment. But all in all, people living in vulnerable areas have had a positive (although modest) growth in employment – ever more people have found employment in other municipalities (Appendix 3, Table 3.3).

4 Localization and Commuting

Looking at all agricultural municipalities in Denmark (including well over 850,000 inhabitants) the income per capita has risen markedly above the national average. For people living in these agricultural municipalities the employment has even risen more than the national average, although the number of local jobs has fallen a little. In a small densely populated country with reasonable road systems it is therefore possible to maintain the employment of most people living in rural areas. They just have to commute to work in other municipalities. But that is not easy on Samsø or the other islands nor on West Lolland where the distances to economic centres may be too long or the problems too heavy.

Besides, it is conspicuous that the growth of jobs is concentrated in relatively few municipalities while employment (i.e. the number of inhabitants employed) has grown in almost all municipalities except the islands and Copenhagen.

More explanations might be found regarding the increased commuting: improved infrastructure and fleet of cars, changed localization of trades and industries and a housing market approaching its saturation point on a countrywide basis. This as well as the significant differences in property prices does not make it very attractive to move to the cities, cf. Figure 3.

Figure 3 Factor income and land values in housing areas in different municipalities. 1988

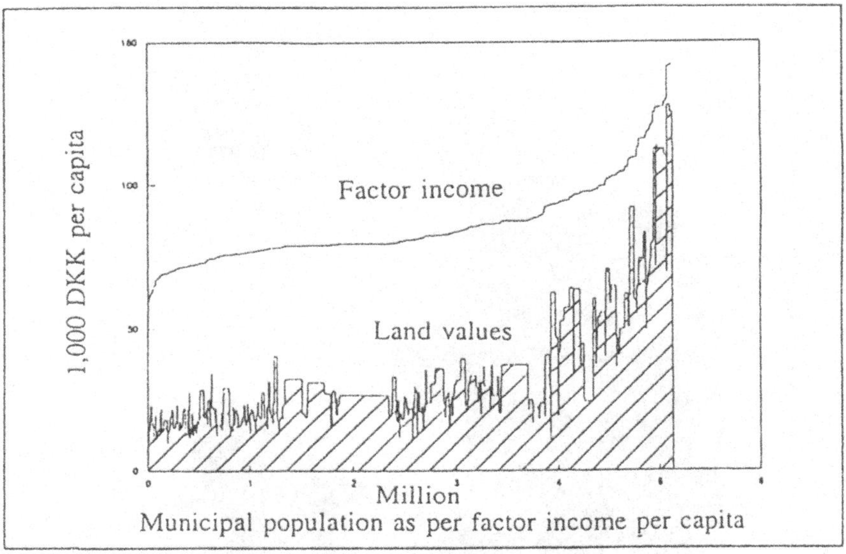

Source: A.C. Hansen »The Geographic Redistribution«, AKF 1991.

Note: On the horizontal axis the municipalities are placed from left to right according to their factor income per capita. They are dimensioned vertically in proportion to their population. Thus, a big municipality indicates greater weight than a smaller one. E.g. the Municipality of Copenhagen has the longest horizontal line. The area below the curve represents the total income.

Figure 4 The growth of local jobs. 1984-1992

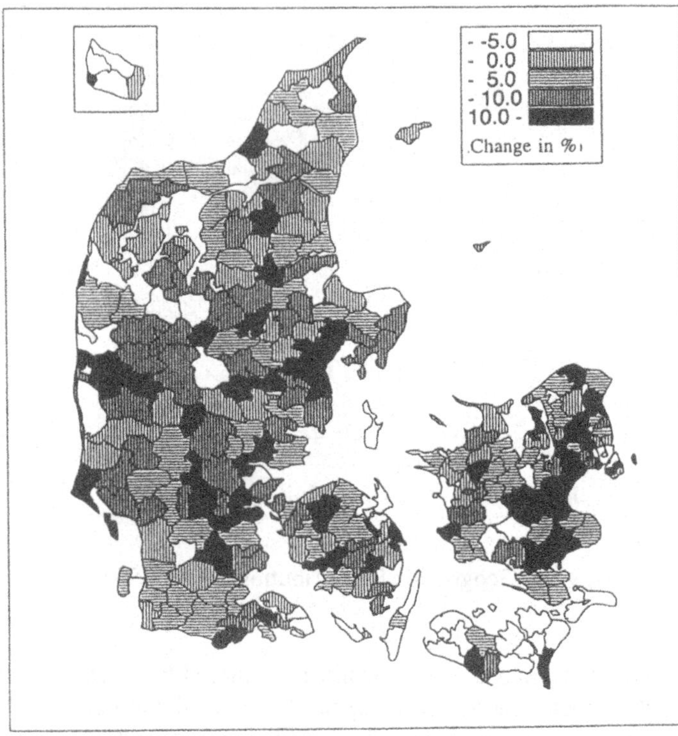

Source: AKF Nyt, August 1994.

Note: The figure shows the percentage growth in the number of jobs in each municipality.

Figure 5 The growth of employment as per residence in 1984-1992

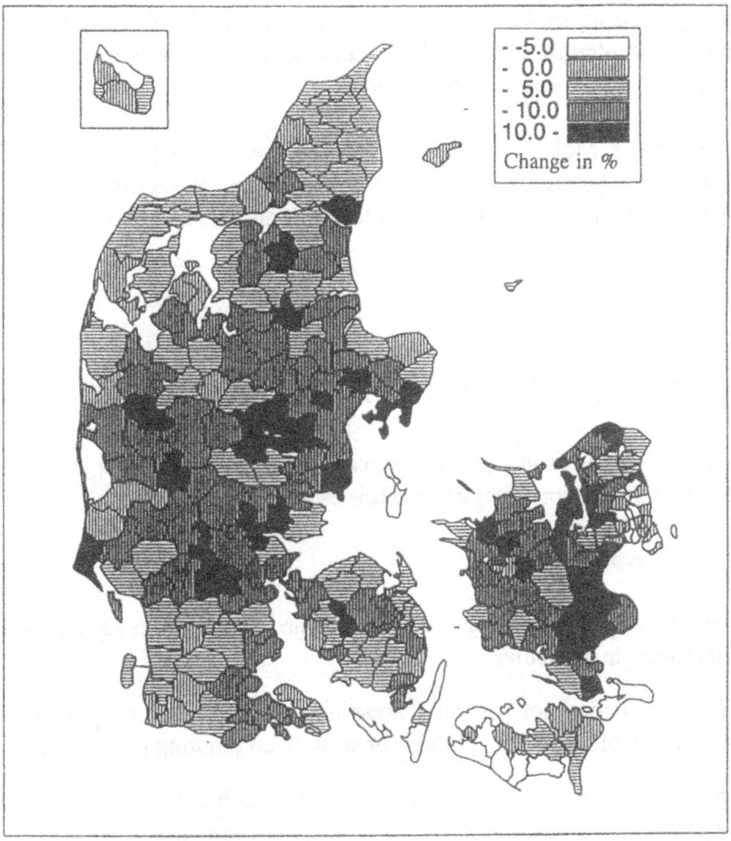

Source: AKF Nyt, August 1994.

Note: The figure shows the growth of employment among inhabitants in each municipality.

5 Concentration of Trade and Industry

It is difficult to imagine that the number of jobs in peripheral areas will grow in the future. The traditional industries – agriculture and fisheries – will continue to employ less and the service industry will to an increasing extent be concentrated in the cities. Industry, too, might very well be more concentrated in selected parts of the country.

As pointed out by Maskell (1994) enterprises within certain trades tend to gather in certain areas. The clothing industry is concentrated in the Herning-Ikast area, where there is a long tradition in that field. Conversely, many companies within the same industry have disappeared from the rest of the country. The explanation might be a geographic concentration of know-how, contacts and labour with relevant training and education – spatial economies of scale.

On the other hand, Molle (1995) has shown that wealth levels in the regions of EU have continuously converged and spatial concentration of industries has continually decreased over the period 1950-90. The textile industry is an exception to this general pattern.

6 Regression Analysis of Geographic Patterns

In Appendix 1 we seek some general patterns for economic growth in different parts of Denmark. A regression analysis model is established to do so.

There are four hypotheses:

- the decline of jobs and population has hit the rural areas in general and the peripheral areas in particular
- the decline of local jobs in rural areas has not to the same extent reduced the number of employed living in those areas with commuting possibilities
- the relative decline in factor incomes in rural areas has not led to a relative fall in real disposable income due to growth in transfers (and taxes)
- the spatial growth pattern might reveal economies of scale (spatial concentration).

The first three hypotheses appear to be confirmed.

Economies of scale appear to be obvious in Appendix 1, Table 1.2. Regression analyses show, however, that if due regard is taken to the special problems of rural and peripheral areas, there are no signs of economies of scale (Appendix 1, Table 1.3). Quite in line with Molle (1995) we find no signs of spatial concentration. But municipalities with high gross income (in 1984) appear to have attracted more inhabitants and employees (in 1984-92) than the rest. On the other hand, disposable income per capita in these growth areas has increased relatively little – presumably another sign of increasing spatial redistribution during the period, but it would also be due to demographic explanations.

The Capital. It is probably in the metropolitan area that Denmark has its most serious regional problems. The growth in disposable income in the Municipality of Copenhagen and the County of Copenhagen is half the growth in the rest of the country. But the low growth is particularly due to a fall in population.

It is no longer only in the Municipality of Copenhagen that the shoe is pinching. A number of residential suburbs has just as miserable an income development, if not worse. Besides a fall in population, the level of employment has dropped and there has been only a modest improvement in per capita income. This depressing trend is due partly to demographic factors: some residential suburbs have simply become »silver-wedding districts« or retirement areas, where people retire from the labour market and in some cases lose their spouse or partner – and remain in the flat or house that was once the framework for a whole family.

So while the Municipality of Copenhagen now has fewer old-age pensioners and a low growth in social transfers (although from a high level), the County of Copenhagen has experienced high growth (from a low level – in 1992 the county still lay one quarter below the Municipality of Copenhagen with respect to social transfers per capita). The percentage growth in transfers in the entire metropolitan area is the same as in the rest of the country, but measured in DKK it is lower, and measured as a share of gross income, lower still. Conversely, the percentage growth of taxes is lower in the metropolitan area than in the rest of the country – simply because of a smaller growth of taxable income in the capital. However, when we measure taxation in DKK we find a distinctly higher rise in the metropolitan region (20% higher than elsewhere) and also find that tax has taken a larger part of a lower income growth.

In view of the depressing trend in Copenhagen, one might ask whether large sums of money should continue to be transferred from the capital to the rest of the country (DKK 18 billion in 1988 according to Hansen, 1991). This is a political choice and part of the welfare system. However, it must not be forgotten that there are greater differences in economic development within the metropolitan area than in the rest of the country, despite big mutual equalizations, and that Denmark's richest municipalities still lie in the metropolitan region.

Figure 6 Employment in selected regions

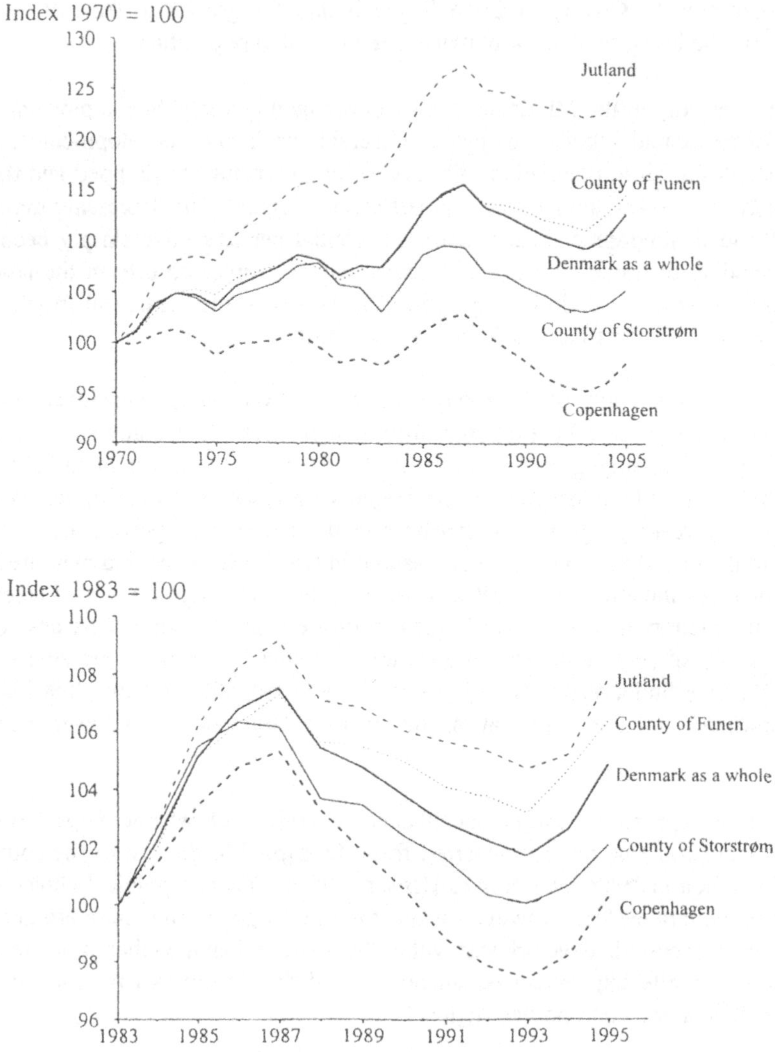

Source: AKF's Data Bank.

Figure 7 Regional differences in unemployment

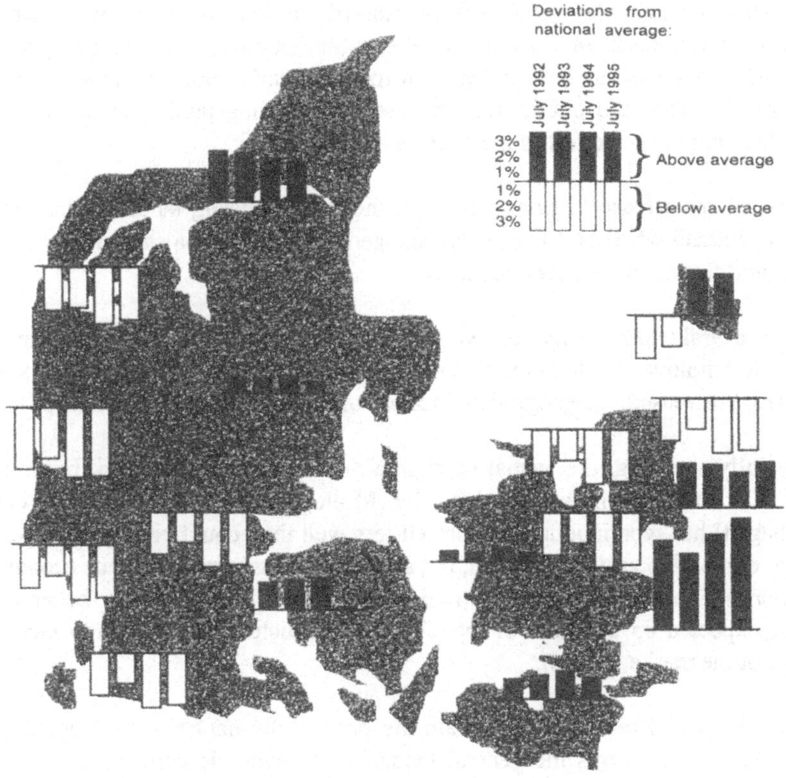

Source: The National Statistical Office of Denmark.

Note: The national average unemployment rate July 1992-July 1995 was 11 %.

Growth Regions and Spatial Redistribution. In contrast to the low growth in the capital, growth has been high in Jutland – and highest of all in the County of Aarhus, where disposable incomes rose one third more than the national average and the level of employment grew two times as much as in the rest of the country. However, the counties of Ribe, Vejle and Nordjylland have also enjoyed high economic growth, and employment has risen quite considerably in both Viborg and Ringkøbing counties (see Appendix 3).

Some of the reasons behind the growth in Jutland and lack of growth in Copenhagen are closely connected to changes in Danish institutions. In 1970, Denmark experienced a reform of local governments, changing responsibilities and activities from the central government (in Copenhagen) to local municipalities and counties. This meant a very considerable spatial redistribution of public activities from the capital to the rest of Denmark. In Copenhagen the public employment rose by approximately 50% from 1970 to 1980, whereas it rose by nearly 100% in Jutland. At the same time the taxes and public transfers rose, channelling public funds from the metropolitan region to the rest of the country.

But there are other reasons. Employment in the manufacturing industries increased steadily in Jutland whereas it fell in Copenhagen. And later on, a similar movement can be found for the private service sector.

There are several explanations: relative wages, land rent and productivity were more attractive to employers in Jutland. And as the home market dwindled and the export market (EU) increased, the geographic location of Jutland became more central.

But generally speaking, the spatial economic development in Denmark resembles that of EU in general, as described by Molle (1995). The former centre (Copenhagen) has continuously performed less well than could be expected, being unable to exploit the growth potential, but also continuously improving its economic base (share). The former peripheral regions in Jutland did repeatedly better than could be expected on the basis of their industrial structure, being able to increase their part of the market (shifts).

So although special reasons can explain the poor performance of the capital, the Danish case also confirms the general tendency to economic convergence among regions.

But the convergence does not include the smaller islands and other isolated municipalities. The spatial barriers appear to be too large for these peripheral areas.

Geographic Distribution and Mobility of Ethnic Minorities. There is a marked contrast between the spatial economic growth pattern in Denmark and the migration pattern of ethnic minorities.

In this paper we distinguish between two categories of ethnic minorities: immigrants and refugees.[1] Nearly all immigrants have come to Denmark during the last 30 years. New immigrants are typically family members of other immigrants, already living in

[1] For a discussion of the distinction and definitions, see Hummelgaard et al. (1995). Immigrants in this paper do not include immigrants from the OECD-countries.

Denmark and will therefore settle in regions with many immigrants. As can be seen in Table 1 nearly 3/4 of all immigrants are concentrated in the metropolitan region whereas very few live in smaller municipalities. 40% of all immigrants live in five municipalities – all in the Copenhagen area. And this pattern remains rather stable.

The refugees have come later than the immigrants. Upon arrival they are scattered all over Denmark. The Danish authorities have established refugee camps in all parts of the country in order to avoid concentration. But as soon as the refugees are allowed to move, they do so.

They move from villages to smaller cities, from there to larger ones and from all parts of Denmark to the metropolitan region, as can be seen in Table 2.

Table 1 The distribution of ethnic minorities and the rest of the population as per municipal groups in 1991, 18-66 year-olds

	Ethnic minorities				Population as whole
	First generation			Second Generation	
	Immigrants	Refugees	Total		
Metropolitan region	72.4	36.7	59.7	76.3	33.9
Copenhagen	30.7	14.5	24.9	33.4	8.8
South-western suburbs	18.1	7.8	14.5	16.0	6.6
Municipalities above 20,000 inhab.	17.2	10.6	14.8	20.5	12.2
Municipalities 10,-20,000 inhab.	5.8	3.2	4.9	5.5	4.6
Municipalities below 10,000 inhab.	0.6	0.6	0.6	1.0	1.6
Outside the metropolitan region	27.7	63.4	40.3	23.7	66.1
Aarhus	5.6	12.6	8.1	5.1	5.5
Odense	3.4	7.9	5.0	3.9	3.7
Municipalities above 40,000 inhab.	7.5	19.0	11.5	5.4	14.7
Municipalities 20,-40,000 inhab.	5.1	10.2	6.9	4.4	9.0
Municipalities 10,-20,000 inhab.	3.4	8.9	5.4	2.8	17.2
Other municip. with more than 33.3% of the population living in towns	1.3	2.8	1.8	1.0	7.2
Other municip. with less than 33.3% of the population living in towns	1.4	2.0	1.6	1.0	8.9
Total	100	100	100	100	100

Source: Hummelgaard et al., 1995.

Table 2 Net migration from 1990 to 1991 of refugees in 1990, per cent as per number

Metropolitan region	3.5
Copenhagen	2.1
South-western suburbs	6.5
Municipalities above 20,000 inhabitants	5.2
Municipalities 10,-20,000 inhabitants	-2.5
Other municipalities[1]	-
Outside the metropolitan region	-4.1
Aarhus	7.4
Odense	0.6
Municipalities above 40,000 inhabitants	-3.1
Municipalities 20,-40,000 inhabitants	-5.2
Municipalities 10,-20,000 inhabitants	-7.1
Municipalities below 10,000 inhab. with more than 33.3% of the population living in towns	-8.5
Municipalities below 10,000 inhab. with less than 33.3% of the population living in towns	-6.6

Source: Hummelgaard et al., 1995.

As shown by Hummelgaard et al. (1995) this pattern in the movements is fairly stable and will eventually change the spatial distribution of the refugees, so that it resembles that of the immigrants.

One might well ask why the ethnic minorities tend to concentrate in the metropolitan region, where demand for labour has grown so little and unemployment is high (as shown in figure 6 and 7). In other countries (US) immigrants are very mobile and tend also to move to regions with growing demand for labour. In Denmark the immigrants and refugees may be mobile when they arrive in the country, but the movements appear to be in the opposite direction: away from the regions in growth towards the region in stagnation.

Some exceptions to the rule should, however, be mentioned. There has been a net migration to two of the counties in Jutland with relatively high growth: Aarhus and Ringkøbing. And mobility is somewhat higher for unemployed than employed refugees.

But far more important than the possibilities of getting a job appears to be the housing situation. Refugees living in municipalities with many (social) flats (as compared to owner-occupied housing) are considerably less mobile than the rest.

Municipalities dominated by social housing are concentrated in the Copenhagen area. And quite a number of flats in this social housing are relatively new, large and expensive. So ordinary wage earners cannot afford them, but families on social allowance get extra subsidies to pay the rent. In this way (jobless) ethnic minorities have come to dominate a number of such housing areas.

This development has of course also a cultural aspect. Most minorities prefer to live together in order to share their common values, culture and background. So a given geographic pattern has a tendency to reinforce itself. But the housing policy and the high standard of social allowances have given little encouragement for new minority centres to develop outside Copenhagen (and Aarhus).

The concentration of ethnic minorities in the metropolitan region has only been documented in 1984-90. But also in this period, the economic growth has been remarkably higher in the western parts of Denmark than in the eastern, as shown in figure 6. During the last years' economic upswing in Denmark (1993-95) there has been an ever falling unemployment in Jutland whereas unemployment in Copenhagen has remained relatively high. There is a general concern that the present growth in exports and production should be hampered by bottlenecks on the labour market in parts of Jutland. And the export industry in Denmark is characterized by its use of unskilled male labour – like most immigrants and refugees.

So the welfare state in general and the housing policy in particular have created problems for the proper functioning of the Danish economy. But of course this is not only a lack of mobility of ethnic minorities. It is just as well a lack of mobility of the Danish labour force in the Copenhagen area and other stagnating regions.

As shown by Mogensen et al. (1995) there is little economic incentive to go to work for more than 20% of the employed in Denmark. Economically they would not be worse off, if they received social transfers instead of wages. So the incentive to move in order to get a job is small for many unskilled unemployed. This could explain the lack of spatial mobility on the Danish labour market, especially among low-wage earners.

7 Conclusion

An ever more generous welfare state has increased the equality in Denmark in the last decades. This is reflected in the spatial economy. Rural areas have benefited from the redistribution at the cost of rich municipalities.

Some of these municipalities are situated in the metropolitan region. The most serious regional problem in Denmark is nevertheless connected to the situation in Copenhagen, with slow growth, high unemployment and considerable social problems.

This raises also the question why the ethnic minorities have been concentrated in and around Copenhagen. Why refugees move from small to larger towns, from Jutland (with job opportunities) to Copenhagen (with none). We do have explanations: the lack of incentives to work in the Danish welfare system. If equality is high on the agenda, it is of course difficult to create more mobility. But considerations for the unemployed, the national economy and the Danish welfare state require a change of politics so that mobility on the Danish labour market is increased.

Regardless of what one thinks about the Danish welfare system there can be little doubt that regional policy in the Kingdom of Denmark is not in the hand of either the Minister of the Environment and Planning or the Minister of the Interior. Taxes, social transfers etc have greater effects on the economic map of Denmark than any intended regional development policy we have seen.

Appendix 1 Regression Analysis of Geographic Patterns

Introduction. In this appendix we seek some general patterns for economic growth and decline in different parts of Denmark.

There are four hypotheses:
- The decline of employment opportunities and population has hit the agricultural municipalities in general and the peripheral areas in particular
- The decline of local employment has not reduced the number of employed living in those rural municipalities where it is possible to commute to other municipalities with a growing demand for labour
- In the rural municipalities the relative decline of factor incomes and jobs has not led to a relative fall in the disposable real income per capita, not least because of the growth in transfers (and taxes)
- The growth pattern among municipalities might have to do with spatial concentration.

1. Rural and Peripheral Areas

The reason for the decline of activity in rural areas is the falling employment within agriculture itself and the derived effects on factor incomes and employment in other local industries.

The rapid growing productivity in agriculture and falling relative prices on agricultural goods should lead to bigger production (measured in quantities), but relatively lower factor incomes (measured in current prices) in agricultural municipalities. For the same reason, local jobs, employment and population are supposed to decrease.

Contrary to this, you have the possibility of commuting to municipalities where the number of jobs are growing.

Add to this that transfers and taxes have risen markedly more than the factor incomes in the period 1984-1992.

Tables 1.1 and 1.2 verify these points quite well. As seen from Table 1.1, in 1984 the small municipalities are characterized by a relatively high number of people employed within agriculture and a relatively low disposable income. Table 1.2 shows that the average gross income in fixed prices has risen most within municipalities below 15,000 inhabitants and least in big municipalities (above 30,000 inhabitants). Contrary to this, the gross incomes in current prices grew least in small municipalities and relatively much in big ones. That corresponds quite well with the fact that big municipalities – where service trades play an important role – have had quite a good growth in income and jobs, while smaller municipalities have had a bigger growth in production, measured in quantities produced.

Table 1.1 Disposable income and employment within agriculture in municipalities distributed on size

	Municipalities[1] grouped by number of inhabitants				
	0 - 5,000	5, - 10,000	10, - 15,000	15, - 30,000	30,000 -
x_1: Disposable income[2] per capita in 1984 (DKK)	36,554	37,846	38,367	38,161	39,090
x_2: Share of people employed within agriculture in 1984 (per cent)	28,5	25,4	21,9	15,1	7,4

Source: Municipal Statistical Data Bank, National Statistical Office of Denmark and own calculations.

1 Exclusive of the metropolitan area.

2 1980-prices.

There also seems to be a connection between the development of employment, jobs, number of inhabitants and the size of the municipality (in 1984). Anyhow, the biggest municipalities have on average the biggest growth, while the smallest ones have the lowest. It is worth noting that even the smallest municipalities on average have a positive employment development (conversely, in the same period the metropolitan area had a decline). In municipalities with 5,-10,000 inhabitants you also see that the number of jobs has only risen by 2.4% while the number of employed living in the very same municipalities has risen by more than 5.0%, thus markedly above national average. Conversely, the big municipalities (above 30,000 inhabitants) have a growth in jobs of 7.5%, where employment has risen by 6.6%. On average more people have commuted from small to big municipalities.

Table 1.2 also shows that the average disposable income per capita has risen most in the smallest municipalities and less in the biggest ones. It is part of the picture that the disposable income has risen considerably less in the Capital than in the rest of the country. The redistribution from rich to poor parts of Denmark seems to have increased markedly in the period in question.

Table 1.2 Growth in population, employment and incomes 1984-1992

	Municipalities[1] as per number of inhabitants in 1984					Denmark
	0 - 5,000	5, - 10,000	10, - 15,000	15, - 30,000	30,000 -	as a whole[2]
Number of municipalities	19	111	45	32	18	275
Number of inhabitants as per municipal group	80,401	817,559	535,445	659,453	1,291,-975	5,112-,005
	Average growth in per cent 1984-1992					
y_1 Local jobs	0.7	2.4	5.2	4.5	7.5	3.4
y_2 Employment (as per residence)	0.6	5.1	5.9	4.6	6.6	3.4
y_3 Disposable real income as per inhabitant	22.0	20.0	20.1	18.3	16.3	16.3
y_4 Gross income[3] in fixed prices[4]	15.0	17.7	17.6	14.3	7.9	11.3
y_5 Gross income in current prices	50.0	54.8	56.8	55.1	56.5	53.4
y_6 Number of inhabitants in the municipality	-3.3	0.5	1.2	1.1	3.3	1.0

Source: Municipal Statistical Data Bank, National Statistical Office of Denmark and own calculations.

1 Exclusive of the metropolitan area.
2 Inclusive of the metropolitan area.
3 Gross income defined as gross tax income before tax deduction (and exclusive of taxes).
4 Deflated with GDP deflator per industry so that the growth describes the change in quantity.

2. Regression Analyses: a Simple Model

In order to test the hypotheses mentioned we have carried out six independent regression analyses. Separately, each of the six variables: y_1 - y_6 forms a response variable in the same regression model. The explanatory variables of the model consist of two continuous variables and two sets of categorized variables. The model is:

$$y_i = \beta_0+\beta_1x_1+\beta_2x_2+\beta_3x_3+\beta_4x_4+\beta_5x_5+\beta_6x_6+\beta_7x_7+e_i$$

$i=1,...,6$

The regression analysis includes 225 municipalities (i.e. all Danish municipalities exclusive of the metropolitan area).

Equation (1) represents thus six regression equations not to be understood as a system, but as six independent regressions.

In succession the response variables have the following definitions:

y_1 Percentage change in the number of local jobs from 1984-1992
y_2 Percentage change in employment (as per residence) from 1984-1992
y_3 Percentage change in disposable real income per capita from 1984-1992
y_4 Percentage change in gross income[2] in fixed prices from 1984-1992
y_5 Percentage change in gross income in current prices from 1984-1992
y_6 Percentage change in number of inhabitants from 1984-1992

The two continuous explanatory variables of the model are:

x_1 The gross income per capita in 1984
x_2 The share of people employed within agriculture in 1984.[3]

The variables $x_3,...,x_6$ represent the categorized variables of the inhabitants in a municipality in 1984.

Basis category: Municipalities with 10,-15,000 inhabitants.

x_3 Municipalities below 5,000 inhabitants
x_4 Municipalities with 5,-10,000 inhabitants
x_5 Municipalities with 15,-30,000 inhabitants
x_6 Municipalities above 30,000 inhabitants
x_7 The variable x_7 represents the categorized variable of peripheral municipalities defined in the following way:

Definition: a municipality is regarded a *peripheral municipality* if the biggest town of the municipality has less than 3,000 inhabitants, if the distance from the municipality to the nearest strong geographical centre is more than 40 kilometres and if the distance to the nearest geographical centre additionally is more than 30 kilometres.

---

[2] The gross income being defined as gross tax income before tax deduction (exclusive of taxes).

[3] The share of people employed within agriculture is defined as the number of self-employed and employees within agriculture in relation to people employed in the total private sector.

Besides, all minor island-municipalities, except Fanø, have been regarded as peripheral municipalities.

In order to construct the categorized variable for peripheral municipalities it has been necessary first to define the geographical centres.

A *strong centre* is characterized by the following criteria:
- more than *40,000 jobs* in the municipality in 1984 and
- a *commuter intensity* > *2.0* in 1984

where the commuter intensity is defined as number of in-commuters in relation to out-commuters.

A *centre* is characterized by the following criteria:

- more than *20,000 jobs* in the municipality in 1984

or
- more than *10,000 jobs* in the municipality in 1984 and
- a *commuter intensity* > *2.0* in 1984.

Centres and peripheral municipalities are shown in Figure 1.1.

Equations 6 and 4 (explaining number of inhabitants and gross income in fixed prices) are estimated by OLS. The observations are unweighted as there are no signs of heteroscedasticity, i.e. we assume that e_i is homoscedastic (for i=6 and i=4 respectively). However, in equation 1, 2, 3 and 5 there is significant heteroscedasticity such that the variances of the errors are larger for municipalities with a small number of inhabitants. Therefore, these four equations are estimated by weighted least squares with the weights depending on the number of inhabitants (n) in each municipality. In equations 2, 3 and 5 the weights are estimated assuming multiplicative heteroscedasticity, cf. Harvey (1976), i.e. $\delta_i^2 = \exp(\alpha_o + \alpha_{1i} \log n)$, when δ_i^2 is the variance of e_i, i=2, 3, 5. Using the same method for equation 1, we estimated an unreliably large value of α_1 (numerically). In equation 1, therefore, α_1 was set equal to -1 (i.e. the observations were weighted by $\sqrt{n}$). The outcome in equation 1 was, however, not sensible to the size of the weights.

Figure 1.1 Regional centres and peripheral municipalities, 1984

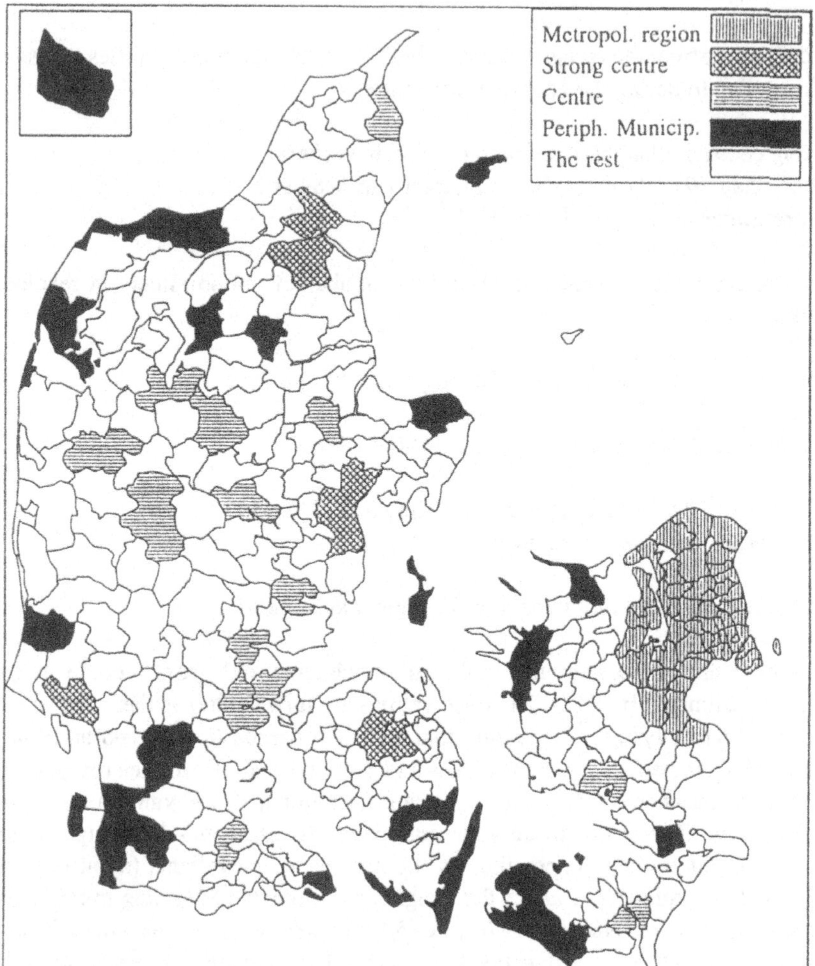

Source: AKF's data base

3. Model Results

The results from the six regression analyses are shown in Table 1.3.

Firstly, it appears from Table 1.3 that *peripheral municipalities have done badly*. This applies to both development in employment and jobs and factor income in fixed and current prices as well as the number of inhabitants. Only the sign of the coefficient for the disposable income per capita is not significantly negative (confirming the importance of the welfare system). The peripheral municipalities – as defined by us – have not had much to write home about in 1984-1992. The hypothesis that it would be difficult to find growth potentials in very many peripheral municipalities seems to have been confirmed.

Secondly, the hypotheses of the development in *agricultural municipalities* are confirmed. The quantities produced have risen significantly more in municipalities where x_2: the share of people employed within agriculture is high. Conversely, the gross income in current prices has risen significantly less. Furthermore, the employment and number of inhabitants and jobs have risen significantly less than in other Danish municipalities (outside the capital). It is, however, remarkable that the negative coefficient for x_2 is three times bigger numerically by explaining growth of local jobs than growth of employment. Comparing two municipalities where one (rural municipality) has 30% of the privately employed in agriculture, while the other one (urban municipality) has 10%, then the regression results suggest that the rural municipality – ceteris paribus – has had 6 percentage points lower growth in the number of jobs than the urban municipality, but only 2 percentage points lower growth in the number of inhabitants employed. Thus it is confirmed that *commuting* plays an increasingly important role in the rural areas. Eventually, the disposable real income has risen more in municipalities where agriculture carries great weight. Even if this result is not significant there is a marked difference compared to the development of the gross income in current prices. The result emphasizes the importance of taxes and transfers to the income development in rural districts.

Thirdly, there seems to be a clear tendency that *municipalities with high incomes* per capita in 1984 also have had high growth in employment, number of inhabitants and gross income in fixed prices, but a relatively low growth in disposable income per capita. If for instance a municipality has had a gross income 10,000 DKK higher per capita than another municipality, then the number of inhabitants would have risen 5 percentage points more and employment (by residence) also 5 percentage points more, while the disposable real income per capita would have risen 5 percentage points less than the other municipality. However, one cannot claim that high-income municipalities have generated any particularly high independent growth. The number of local jobs has risen more in municipalities with high gross incomes than in other municipalities, but not significantly so in this analysis. It could probably be a question of people moving to these municipalities without necessarily getting work there. E.g., it might be a question of suburban municipalities in connection with big provincial cities. But generally speaking it is easier to explain the spatial development than that of job creation.

Table 1.3 Growth of population, employment and income in municipalities with different income, size and locality - coefficients for six regression analyses

Explanatory variables:	Response variable (percentage change 1984-1992)						
	y_1: Local jobs	y_2: Employment as per residence	y_3: Disposable real income per capita	y_4: Gross income in fixed prices	y_5: Gross income in current prices	y_6: Number of inhabitants	Average variable unweighted
Intercept	-8.17 (0.62)	-20.75 (-2.89)	46.79 (5.77)	-15.51 (-1.78)	49.97 (4.70)	-26.98 (-5.56)	
x_1: Gross income in 1984 per capita	0.0003 (1.60)	0.0005 (4.24)	-0.0005 (-3.69)	0.0005 (3.30)	0.0002 (1.11)	0.0005 (6.49)	57,310 DKK[1]
x_2: Share of employed within agriculture in 1984	-0.29 (-3.90)	-0.09 (-2.37)	0.07 (1.51)	0.28 (5.81)	-0.19 (-3.27)	-0.07 (-2.51)	22.1%
x_3: Municipalities below 5,000 inhabitants	-0.72 (-0.23)	--0.46 (-0.30)	-1.95 (-1.36)	-0.9 (-0.47)	-1.72 (-0.81)	-1.24 (1.40)	
x_4: Municipalities with 5,-10,000 inhabitants	-1.14 (-0.78)	0.65 (0.85)	-0.6 (-0.71)	0.32 (0.35)	-0.14 (-0.13)	0.26 (0.49)	
x_5: Municipalities with 15,-30,000 inhabitants	-2.02 (-1.30)	-2.37 (-2.73)	-1.85 (-0.78)	-1.49 (-1.50)	-2.67 (-2.11)	-1.08 (-1.53)	
x_6: Municipalities above 30,000 inhabitants	-3.34 (-1.98)	-2.25 (-2.33)	-1.35 (-0.95)	-2.24 (-2.04)	-4.1 (-2.87)	-1.06 (-1.17)	
x_7 Peripheral municipalities	-4.15 (-2.25)	-5.05 (-5.40)	-0.79 (-0.85)	-4.39 (-3.79)	-5.7 (-4.26)	-2.77 (-4.65)	
Average (unweighted)	3.7%	4.9%	19.7%	16.1%	54.9%	0.54%	

Note: The gross income is all taxable incomes before tax deductions and taxes. The figures in brackets are t-values.

[1] 1980-prices

Fourthly, Table 1.3 seems to dismiss the hypothesis of economies of scale. When taking the special problems of the peripheral areas into account as well as the importance of agriculture and income levels in 1984, the size of the municipalities cannot deliver any isolated explanation of the development in income, employment or number of inhabitants. If there is any systematic difference, it would rather point towards diseconomies of scale. We have also tried to estimate equations with the size of the municipality as a continuous variable. The coefficient for such a municipal size became significantly negative. We did, however, prefer the categorized size variables, as there appear to be no clear size function (as shown in Table 1.3 it is certainly not linear and it differs from equation to equation). For the time being, nothing shows that big urban communities generate a higher growth regarding the number of jobs than smaller ones – taking the facts mentioned into consideration.

Thus it is clear that a basis of further investigations exists where the factor income is more clearly defined and is distributed by industries. In this way it should be possible to go deeper into the problems of the spatial concentration of industries.

Appendix 2 The Public Sector and Public Transfers

During the 1960s and 1970s a big expansion of the public sector took place. However, in 1982 the development culminated and the public sector's share of the GNP dropped till 1990 after which some modest increases within the relative shares have taken place.

Table 2.1 Public expenditure in 1975 and 1993

	1975	1993
	Percentage of GNP	
Public spending	24.6	26.3
Hereof:		
– administration	2.1	2.2
– judicial system and defence	3.5	3.1
– welfare services	17.2	18.8
– economic services etc	1.8	2.3
Investments (net)	3.7	2.0
Transfers, subsidies etc	19.1	28.1
Interest	1.2	7.7
Total expenditures	48.6	64.1

Source: Financial Statement, August 1994.

Contrary to this there has still been an increase in the current transfers including household transfers, cf Table 2.1 and Figure 2.1.

Figure 2.1 Development of current transfers 1971-1993

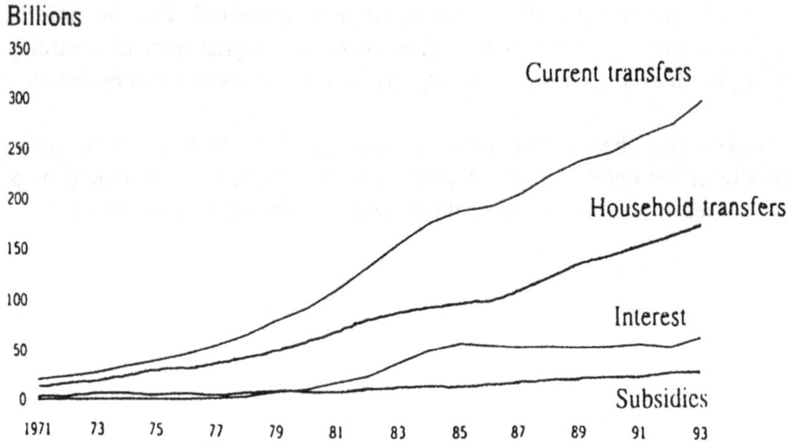

Source: Statistical Ten-year Review, 1994.

As seen in Table 2.2 the average real growth in transfers per year was 4.4% from 1973 to 1993. The Statistical Ten-year Review (1994) calls it »quite a considerable growth compared to the fact that the total economy grew by 1.8% per year in the same period. The real improvement of transfers is a result of the following conditions:

- The number of persons receiving transfer incomes has grown at the same time as the average payment period has been expanded for some of the transfer incomes. This goes especially for cash benefits and unemployment benefits for the period 1972-1982 where the crisis referred many new people to transfers.

- A real improvement of the level of some transfer areas has taken place, e.g. pensions and benefits to families with children.

- Expenditure-demanding reforms have been carried out in several transfer areas including the initiation of new types of transfers. In the first part of the period this goes for child benefit etc, while in the second part a rise in education grants and housing subsidies has taken place.

- More people than before use the transfer system.«

Table 2.2 Income transfers to households 1972-1993

1993 prices	1972	1982	1993	Real growth		
				1972 -82	1982 -93	1972 -93
	Million DKK			Per year		
Total transfers	72,494	128,264	178,921	5.87	3.07	4.40
Social pensions	38,291	47,026	61,218	2.08	2.43	2.26
Civil servants retirement payments	5,324	8,035	10,823	4.20	2.74	3.44
Early retirement benefit	*	8,501	12,258	*	3.38	*
Unemployment benefit	3,331	28,777	35,757	24.06	1.99	11.97
Cash benefit	3,170	9,106	12,095	11.13	2.61	6.58
Sickness and maternity benefit	5,082	8,067	10,365	4.73	2.30	3.45
Child benefit etc	2,303	7,369	12,581	12.33	4.98	8.42
Housing subsidies	1,984	2,967	7,321	4.11	8.56	6.41
Educational grant etc	2,197	3,115	5,675	3.56	5.60	4.62
Other transfers	10,813	5,300	10,828	-6.68	6.71	0.01
Real growth of GNP				1.66	1.99	1.83

Source: Statistical Ten-year Review, 1994.

Table 2.3 shows the redistribution in 1987. Households with an income below 200,000 DKK receive more public benefits than they contribute. The consumption potential for this group (about 1 million households) is bigger than they are able to pay out of their factor income. Conversely, about 1.2 million households with an income above 200,000 DKK contributes to the redistribution. According to the National Statistical Office of Denmark »the statistics indicate that an actual redistribution from higher to lower incomes takes place«.

Table 2.3 Redistribution through the public sector

	Total				
	Below 100,000	100,000 - 199,999	200,000 - 299,999	300,000- 399,999	400,000 and above
	DKK per household				
Citizen payments to public authorities	22,136	65,582	111,272	153,034	232,130
Direct taxes	7,925	42,919	77,226	106,474	172,121
Indirect taxes	14,211	22,663	34,046	46,560	60,009
Public services to the citizens	72,906	68,827	61,876	66,729	64,116
Transfers	51,300	48,671	35,273	26,501	17,314
Selected grants/subsidies	21,606	20,156	26,603	40,228	46,802
Net transfers	50,770	3,245	-49,396	-86,305	-168,014
	In thousand				
Number of households	469	518	436	417	363

Source: Statistical Statement, 1994.

In Statistical Statement 1994 is written: »There is a comprehensive division of labour within the public sector due to certain basic principles of public activity:

- the public sector should make task solutions as close to the citizens as possible
- the financing of the public sector – burden of distribution – should be a joint concern.

The ideal goal is to make the public service level and the tax rates go together. If the municipal tax rates and the municipal service level are going to be connected, differences in political decided expenses for e.g. educational or social purposes must be neutralized. It is not reasonable if a municipality with many old people has to have a high tax rate because of the low tax base that pensioners traditionally provide at the same time as having large expenses for old-age pensions the payment of which is compulsory. In order to secure a reasonable correlation between task and burden big transferrals among public sub-sectors take place.«

Appendix 3

Table 3.1 Growth of income, population and jobs in some Danish regions. 1984-1992

	Percentage growth in				
	total disposable income	disposable income per capita	number of inhabitants	number of jobs in the region	employment residence)[1]
Municipality of Copenhagen	8.8	13.1	-3.8	-13.4	-3.7
County of Copenhagen	10.0	12.3	-2.1	8.8	-5.9
County of Frederiksborg	18.5	14.4	3.5	9.9	6.4
County of Roskilde	22.1	14.9	6.3	10.0	10.7
County of Vestsjælland	18.2	15.2	2.6	4.2	6.5
County of Storstrøm	14.1	14.3	-0.2	0.1	2.4
County of Bornholm	8.5	12.5	-3.6	1.0	-2.1
County of Funen	16.7	14.3	2.1	4.3	4.8
County of Sønderjylland	16.5	16.1	0.4	4.6	4.1
County of Ribe	19.8	17.3	2.1	8.1	6.7
County of Vejle	19.2	17.1	1.8	7.7	7.7
County of Ringkøbing	18.6	16.7	1.6	5.4	5.6
County of Aarhus	21.8	16.6	4.4	8.3	8.7
County of Viborg	18.8	19.3	-0.5	6.8	3.9
County of Nordjylland	17.2	16.3	0.8	4.2	4.4
The metropolitan area	13.1	13.6	-0.5	-.03	-0.6
The rest of Denmark	18.2	16.2	1.7	5.5	5.6
Denmark as a whole	16.3	15.2	1.0	3.4	3.4

Source: AKF's Local Accounts.

1 Employees living in the region.

Table 3.2 Growth of income from 1984-1992

	Percentage growth in		
	disposable income per capita[1]	total disposable income[1]	total gross income[1]
Vulnerable municipalities			
– lowest group	15.7	11.7	11.3
– medium group	15.9	14.8	16.0
– highest group	15.6	16.0	17.0
– total	15.8	13.4	13.9
West Lolland	14.7	6.3	4.3
Big islands	14.8	9.6	10.2
Fishery Municipalities	15.4	13.4	11.8
Agricultural Municipalities	17.1	16.4	17.2
Municipalities dependent upon agriculture	17.0	15.9	16.3
Five big cities	14.3	19.0	19.8
The metropolitan area	13.7	13.1	14.2
The rest of Denmark	16.3	18.3	19.2
Denmark as a whole	15.3	16.4	17.3

Source: AKF's Local Accounts.

1 The income has been deflated by a retail price index and thus computed in 1980-prices.

Table 3.3 Growth of population and employment from 1984-1992

	Percentage growth in		
	population	local jobs	employment, residence
Vulnerable municipalities			
– lowest group	-3.4	-2.3	-0.4
– medium group	-1.0	-1.4	1.1
– highest group	0.3	3.3	4.1
– total	-2.1	-1.8	0.5
West Lolland	-7.3	-11.3	-5.1
Big islands	-4.5	-0.3	-2.6
Fishery Municipalities	-1.7	2.3	2.8
Agricultural Municipalities	-0.6	-0.2	3.7
Municipalities dependent on agriculture	-0.9	1.0	2.9
Five big cities	4.1	7.0	6.3
The metropolitan area	-0.5	-0.3	-0.6
The rest of Denmark	1.7	5.5	5.6
Denmark as a whole	1.0	3.4	3.4

Source: AKF's Local Accounts.

References

Articles about inequalities in Danmark written by Hans Hummelgaard & Leif Husted, Bjarne Madsen & Lars Rich and Peter Maskell, AKF Nyt August 1994, AKF Forlaget, Copenhagen.

Christoffersen, Henrik (ed.) (1991), Regionale forskelle i levevilkår, AKF Forlaget, Copenhagen.

Danmarks Statistik (1994), Statistik Tiårsoversigt.

Finansministeriet (1994), Finansredegørelse 94, Copenhagen.

Groes, Nils (1991), Storebæltsregionen, AKF Forlaget, Copenhagen.

Groes, Nils (1994), Uligheder i Danmark, AKF Forlaget, Copenhagen.

Hansen, Anders Christian (1991), Den geografiske omfordeling, AKF Forlaget, Copenhagen.

Harvey, A.C. (1976), »Estimating Regression Models with Multiplicative Heteroscedasticity«, Econometrica, 44, 461-465.

Hummelgaard, Hans et al. (1995), Etniske minoriteter, integration og mobilitet, AKF Forlaget, Copenhagen.

Madsen, Bjarne & Chris Jensen-Butler (1994), Modelling the Regional Economic Consequences of Environmental Policy Instruments Applied to the Transport Sector, Paper presented at the 34th European Congress in Regional Science Association, Groningen.

Madsen, Bjarne & Lars Rich (1994), »Status og udvikling«, AKF Nyt, AKF Forlaget, Copenhagen.

Maskell, Peter (1992), Nyetableringer i industrien, Handelshøjskolens forlag.

Maskell, Peter (1994), »Kager i Vejle, tøj i Herning«, AKF Nyt, AKF Forlaget, Copenhagen.

Mogensen, Gunnar Viby; Søren Brodersen; Lisbeth Pedersen; Peder J. Pedersen; Søren Pedersen & Nina Smith (1995), Hvad driver værket? om sammenhængen mellem socialpolitik, skattelovgivning og arbejdsudbud i dagens Danmark, Rockwool Fondens Forskningsenhed, Spektrum.

Molle, Willem (1995), The Regional Economic Structure of the European Union: an Analysis of Long Term Developments, Paper presented at the Conference on the Occasion of the 25th Anniversary of the Founding of the Institute of Regional Research at the University Kiel.

Mortensen, Jørgen Birk; Bjarne Madsen & Chris Jensen-Butler (ed.): Modelling the Economy and the Environment, Scandinavian Approaches, (Springer Verlag, forthcoming).

Nørskov, Anders (1991), Offentlige og mellemregionale pengestrømme i Danmark i 1988, AKF Forlaget, Copenhagen.

Rowntree Foundation, Joseph (1995), Inquiry into Income and Wealth, York.

Statens Jordbrugsøkonomiske Institut (1993 & 1994), Dansk landbrugs økonomi og fremtidsudsigter, Copenhagen.

Considering Regional Effects of Transport Infrastructure Investments in Privately Financed Projects

Werner Rothengatter
Universität Karlsruhe
Institut für Wirtschaftspolitik
Kollegium am Schloß, Bau IV
D - 76122 Karlsruhe

1 Introduction

During the last decade there has been a resurgence of interest in private financing and increased mobilisation of private capital for transport infrastructure. One reason is fiscal pressure because of a shortage of public funds. Other reasons are the improvement of efficiency of construction and operation, the appropriate treatment of risk and the market conforming allocation of the total costs associated with the project.

In the project finance approaches the usual models applied are the BOT (Build-Operate-Transfer) or the BOOT (Build-Own-Operate-Transfer). The project company, owned usually by construction contractors, equipment suppliers and experienced public utility operators, builds the required facility, operates it long enough to pay back the debt finance raised and remunerate the share-holders, and transfers title, rights and responsibilities of further operation of the facility back to the government. The Channel Tunnel is the "flagship" for a BOOT-project, including a concession period of 65 years. It is the prototype for the workability of private finance and for its limitations as well. The major lesson which can be learnt from this project are that the responsibility and leadership should be with the project company from the very beginning of the planning process and that the roles of the private and public partners should be most clearly defined, the risks properly allocated. This means that setting up the concession contract is one of the most important steps in building the base of a private finance and management.

Defining the roles of the private and public partners properly means that the state should be responsible for controlling the externalities while all internal effects can be operated through market mechanisms. Externalities consist of environmental, safety and spatial effects which are not recognised by the private investors because they are not able to capitalise these benefits. In the case of positive spatial externalities a financial participation of the state is justified, which may consist in capital grants, interest guarantees or risk taking by SOEs (state owned enterprises).

Measurement of regional external economic effects is tedious and its treatment in cost-benefit appraisals often leads to a double counting of benefits as is underlined by the German standardised CBA for federal transport investment. The examples of the Öresund and Fehmarn Belt fixed links show that there is little reason to expect large regional external economies from every megaproject. In developed economies the spatial impacts which are not captured by users and operators might be marginal. This raises the question whether high national or European investment subsidies are justified from the viewpoint of stimulation of external regional growth. There are also good reasons to argue that in the developed European countries the internal growth effects are predominant such that further major extensions of the transport infrastructure should be predominantly a matter of the private willingness to invest.

2 Public and Private Interest in the Transport Infrastructure

Transport infrastructure basically is not a public but a club good. This means that it is provided for a subset of members of the society and that the non-exclusion principle does not hold. Infrastructure being a club good, the club of users in principle should be able to pay the cost of infrastructure provision, maintenance and operation. If the willingness to pay of the users is sufficiently high then there will be a vested interest of private investors to build operate and finance the infrastructure. However presently there seem to be not many infrastructure projects which have financially viable prospects such that relying on the private investors alone would lead to a drastic cut of infrastructure investment plans. However there is some historical evidence and still a strong belief that transport infrastructure will generate a number of external benefits, i.e. benefits which a private supplier is not able to capture. Among those the following effects are often mentioned:

- network synergy, i.e. contribution to a better efficiency of other parts of the network;
- reduction of external costs, such as environmental damage and accident impacts;
- accelerating economic growth in regions affected by the investment; and
- contributing to regional activity issues (improving accessibility to remote regions, reducing the variance regional growth patterns).

The expectation that the latter effects are quite large, potentially dominating the private benefits which can be captured by the market mechanism, has been a dominating motive in the past to supply the infrastructure by the public. This proposition of dominating external versus internal benefits is increasingly being questioned and there is an upcoming scientific reasoning that the private interest in infrastructure projects should be high enough to encourage private investors to take a substantial part of the investment and of the risk.

3 Need for Private Investment Capital

During the previous century, private capital has already played an important role in the development of the infrastructure, at that time in particular for investment in railways. Bankruptcies and a presumed underinvestment by restricting to private finance have moved the public sector into the transport infrastructure provision such that during the most part of this century public financing has become much more common. During the last decade there has been a resurgence of interest in private financing and also in increased mobilisation of private capital for transport infrastructure. The most important reason is increased fiscal pressures on most governments with the resulting shortage of public funds. The situation of the public budget in many countries looks serious. What concerns for example the requirements set for a fiscal stability which is sufficient to join the Monetary Union in Europe (EG: yearly budget deficit lower than 3 % of GDP, total public debt lower than 60 % of GDP) only a few countries show prosepcts to pass the thresholds. But this does not mean that in these countries the investment budgets are sufficient to implement all infrastructure investment plans. For instance in the case of Germany it is expected that only 60 to 70 % of the federal transport investment plan 1992 can be realised until the year 2012 by the usual way of public finance. But beyond the pure fiscal interests there are other efficiency oriented principal aims of private finance initiatives. These are:

- Improve efficiency of provision of public services,
- improve the quality of service provision,
- reduce reliance on public sector finance, and
- reduce overdesign of public sector projects.

So the basic economic idea behind private partnership in infrastructure investments is to considerably improve on the economic viability of the projects.

An indicator for inefficiencies of public infrastructure provision is the remarkable cost overrun with public projects which could be observed in the past. In Germany nobody is surprised to learn that the high speed rail link between Cologne and

Frankfurt is now estimated to cost about 8 Bill. Deutschmarks while the cost estimate for the federal transport investment plan was 4,5 Bill. DM only. For the Great Belt project in Denmark the cost figures have been revised several times, usually accompanied by a revision of the traffic forecast. As the project has been set up by a state own enterprise (SOE) there is no surprise that the revisions of cost and revenue figures point to the same direction (see Figure 1).

Figure 1 Cost overruns for the Great Belt project in Denmark
Source: Flyvbjerg, Bruzelius and Rothengatter, 1995

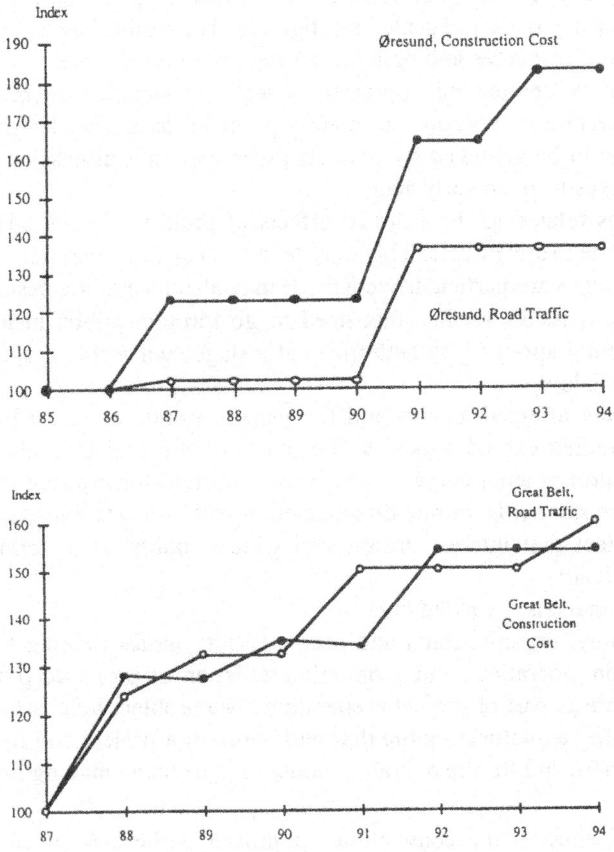

These examples show that there is obviously something wrong with infrastructure investment projects which are publicly financed or for which the public takes all risk.

4 The Conventional Scheme of Planning Large Infrastructure Projects

Traditionally projects have been planned and financed by the public or have been backed by public (providing guarantees). The Channel tunnel is a notable exeption while the Öresund and Great Belt projects correspond to the conventional approach which is described below.

Characteristics and problems of the conventional approach to project development and appraisal are (see Flyvbjerg, Bruzelius and Rothengatter, 1995):

- The project cycle does not include a pre-feasibility phase before a decision is taken to carry out a full scale investigation. The result may be an over- commitment of resources and political prestige at an early stage;
- Project development and appraisal are seen as a technical exercise with a focus on technical solutions at an early point in time. The decision of policy objective to be achieved by projects gives way to a discussion of technical alternatives from an early stage;
- Concerns related to the external effects of projects are not addressed until later in the project cycle. This may lead to project changes at a stage when such changes are particularly costly. It may also lead to the destabilisation of projects as issues surface that need to go through public hearings or that need formal approval by authorities at a stages when this scope for making changes is low;
- Negatively affected groups and the general public are only involved to a rather limited extend and at a late phase of the project cycle. Input from interest groups and general public is underutilised for instance in setting performance standards. Public dissatisfaction with projects may increase by the simple fact that interest groups and general public is underinformed and feels left out;
- No risk analysis is carried out;
- Institutional organisational and accountability issues relating to the implementation, operations and economic regulation of proposed projects play a minor role as part of project preparation. The problem here is lack of definition of the regulatory regime that will apply to a project and its consequences for risks and for the overall appraisal and decision making process.

A major deficiency of the conventional approach is the lack of clearly defined roles for the involved parties. This deficiency is usually not recognised after a short glance at the planning scheme. This is because it appears that the public is acting according to a welfare maximisation goal and trying to control all steps of the process according to social issues, which are following from this general goal setting. So the conventional approach defines the role of the government to be in control of every step of the process, including the choice of technology and mode of implementation, operation and finance. The government does play a variety of roles, some of them in a sense being conflicting. The question that has to be asked is whether the government can act effectively as both a promoter of a project and as the guardian of public interest issues, such as protection of the environment, safety, regional development and finally take the role to safeguard the taxpayer against unnecessary financial risks. Looking at this variety of potentially conflicting roles serious doubts emerge whether the public is able to put these issues into a consistent and workable institutional scheme.

From the viewpoint of institutional economics it is highly probable that private interest groups try to capture public institutions involved to achieve super-normal profits from the infrastructure provision business. In the terms of Robert Stigler this can be called a "rent seeking activity". And the incentives generated by this type of the planning process contributes to maximise rent seeking using all instruments to influence the public decision making. As risk taking of the private sector is limited to the cost risks defined in the terms of trade for the tenders for the construction work the private interest is to

- maximise the size of the project;
- hold many tasks in the terms of reference open and in a fuzzy status such that they are not subject to a clear tendering process;
- define the standards of the technology (technical specification) applied in a way that a limited number of domestic companies are left finally for the tender;
- Allocate the maximal part of the project risk to the public, i.e. to the taxpayer.

5 Alternative Schemes for the Planning Process and for the Investment Appraisal

Usually private financing for transport infrastructure investments is based on BOT (build operate transfer), BOOT (build, own, operate, transfer) schemes or variants of these basic schemes. The Channel-Tunnel is the "flagship" of such a concession financing the transport sector. It is a BOOT-project with the concession originally lasting 54 years and later extended to 65 years as a part of a settlement regarding responsibility for cost overruns. The fact that Euro-Tunnel company is bankrupt with respect to the usual definition of financial bankruptcy does not mean that the BOT-model has proven to fail. A series of lessons can be learnt indeed from the Channel-Tunnel experience, one of them telling that there clearly is a role of public even if infrastructure projects are planned to be financed completely on a private base and that this public role has to be played with a full commitment.

The six basic instruments which are constitutive for an alternative approach aiming at an increased accountability of the total planning process are:

- transparency,
- performance specifications,
- explicit formulation of the regulatory regime,
- clear identification - and where relevant elimination - of policy risks before decisions are taken,
- the involvement of risk capital and
- inclusion of secondary business related to the link.

In this overview only a brief elaboration on the performance specification and the involvement of risk capital will be given.

The use of performance specification implies a goal-driven approach to feasibility studies and decision making.

As far as possible all requirements with respect to a possible infrastructure investments are to be decided before fixing the technical alternatives for providing the infrastructure and before appraising the proposed project. In engineering and also in business economics the performance specification is a usual step which comes first when new products or investment activities are planned. In the case infrastructure projects this means that the specification are derived from policy objectives and public interest requirements to be met by the project, for instance regarding economic performance, environmental sustainablity and safety performance. One advantage of using a performance specification approach is that it forces people to focus on the end rather than on the means, i.e. on questions of what they want to achieve or what to avoid rather than the technical characteristics of a project.

One of the most important issues for accountability of infrastructure projects is a clear allocation of risk. As a high interest of a private sector should be involved when it comes to major projects for the transport infrastructure this implies a substantial risk taking of the private sector. Although one can not give a clear numerical figure which indicates that this requirements is settled a private share of the capital risk of at least 50 % is desirable and the minimum share of a private involvement should not be below about 1/3 of the capital invested. The involvement of private risk capital will insure a higher degree of involvement by the lenders during the final design, construction and operation of the project, and a more effective monitoring. As a consequence, better cost control can be expected and also better controls against construction delays. Recently, the World Bank has shown that it may take several percentage points of interest advance to offset even moderate inefficiencies in terms of cost overruns and delays stemming from the inefficient monitoring of the project (The World Bank, 1994).

What concerns the role of the public it is obvious that

- all activities which involve legal consequences (in particular expropriation of the land owners),
- design and organisation of the public discussion process among all groups involved and interested,
- development of an institutional scheme for the project,
- control of all externalities, in particular
  - consideration of regional externalities

come within the responsibility of the public sector. The economic evaluation of these factors determine the upper limit for a public share in the capital invested.

In the following we discuss how this process may work in the field of regional externalities.

6 Assessment of Regional Effects

6.1 Infrastructure as a Production Factor

It follows from production theory and from empirical evidence that there is a mutual interdependency between economic development and the provision of transport infrastructure. Lack of appropriate transport infrastructure can lead to bottleneck situations for economic growth which is then hampered or diverted to other regions. The relevance of infrastructure for economic growth can be tested by a number of economic approaches. One of them is the questioning of firms and experts with respect to locational choices and their dependence on infrastructure provision. The importance of the single attributes for locational choice is usually evaluated by ranking schemes, as for instance has been carried out by the IFO-institute (1990). While the IFO study comes to a relatively high estimate of the influence of the transport, telecom and energy infrastructure on the competitive power of regions (about 35 % of the total) a PROGNOS study, published in 1990, results in rather modest figures. Contrary to IFO the PROGNOS approach goes more deeply with the sectoral break-down of the economy and in a more detailled classification of types of the infrastructure supply. The basic result is that the regional infrastructure capital is of a high influence for some sectors such as chemical and steel production. But looking at the average over all industrial sectors the relevance of the infrastructure for the regional position in competition seems to be low. So the results of questionnaires show no striking evidence for a strong influence of the infrastructure provision for transport facilities on the relative economic position of regions.

6.2 Evaluation of Regional Benefits of the Transport Infrastructure in the BVWP (Bundesminister für Verkehr, 1993)

The evaluation method for the German federal transport infrastructure investment plan (BVWP) includes spatial effects as an infrinsic element of the overall appraisal of investments. The spatial effects are subdivided by

- regional employment effects induced by construction (NR1)
- regional employment effects induced by operation (NR2)
- effects on the spatial structure (NR3), and
- benefits for the international change of information and trade (NR4).

The NR1 benefits are estimated on the base of an alternative cost approach. It is assumed that the employment effects of infrastructure investment help to reduce the insurance indemnities for unemployed workers. The increase of regional employment is calculated on the base of labour productivities weighted by the share of investment cost which is expected to be spent within the region. The expected weight of unemployment for a region follows from the long-term economic forecast and expected payment for unemployed from a common federal/state taskforce "improvement of the regional economic structure".

The quantification of NR2 is based on the estimation of long-term transport induced employment. It follows from econometric studies that the provision of one kilometer of motorways in West Germany is related to an increase of employment by eight workers. For East Germany is was assumed that even 24 working places can be created by one kilometer of motorway investment. This basic value is weighted by regional preference and other factors. For instance the regional preference factor is composed of an indicator for regional backwardness and an indicator for the regional relevance of infrastructure.

The spatial structure benefits NR3 are derived from the criteria "accessibility" and "economic backwardness". From these criteria a weighting factor is generated which is multiplied by the benefits from the total savings of operation costs, time-costs and regional employment effects. Because of the scaling of the weighting factor the spatial structure benefits can sum up to 30 % of the total monetary benefits for West German regions and 40 % of the East German regions. As the mathematical aggregations scheme for the single benefit items is additive this means that the spatial structure benefits add up and can increase the total monetary benefits by up to 40 %.

Investment projects which contribute to an improvement of international relationships are added a bonus benefit (NR 4) which may sum up to a maximum value of 10 % of the savings of operation and time costs. The relevant indicator for the calculation is the share of international traffic of the total traffic on the link considered.

6.3 Macroeconomic Approach to Estimate the Overall Effects of Infrastructure Investment

While the above mentioned approaches are based on regional cross-section analysis and reflect the change of relative positions of regions which are provided with a good infrastructure supply the macroeconomic approach estimates the overall change of economic productivities stemming from infrastructure investments, usually applying time series analysis.

Aschauer has shown in his publications of 1989 and 1993 that there is a close correlation between the time paths of overall economic productivity and of public investment. Both rising in the 1950_s and 1960_s, and both falling in the 1970_s and 1980_s. On this base Aschauer estimates that the production elasticity of public infrastructure capital is much higher than that of private capital. So he argues that neglecting the extention of public infrastructure in the US since the mid of the seventies has created America_s "third deficit" and has been the most important factor for the slowing down of the general productivity development.

6.4 Critical Remarks

With aspect the work of Aschauer one has first to remember the old lesson that correlation is not implying causality. So while there is little doubt that Aschauer's work is a nice first step which would justify to formulate some hypotheses which would have to be validated by much more detailled approaches, Aschauer himself has devaluated his findings by a political interpretation which cannot be derived from the econometric result. Consequently there has been a heavy criticism debating Aschauer's macroeconomic approach and his implications which has been summarised by Gramlich (1994). The reliability of the Aschauer's figures has also been questioned by professional macroeconomists regarding the implicit outcome of his studies that public infrastructure investments have a cost benefit ratio of more than ten over one and come out much more efficient than private investments.

Also the approach of the German Ministry of Transport has been subject to criticism because it includes biases and double countings. The evaluation method is biased insofar as the evaluation of benefits from time savings is based on the assumption that there is a full employment of ressources, while the evaluation of regional benefits starts from the assumption of underemployed work force. In one of the many subcriteria, the benefits from changing spatial structure (NR 3), these biases are even combined in a way that full employment benefits are weighted by underemployment indicators (regional backwardness). The double counting of effects stems from the characteristics of the German approach which is to transfer multicriteria indicators into monetary scales and to add up the results without checking the overlaps. Consequently the German method tends to overestimate the regional effects of infrastructure. To give an example: For the Maglev link between Hamburg and Berlin the regional benefits were estimated to make one third of the total benefits. But remarkably enough no stations were considered for the Maglev in the economically low developed regions of Mecklenburg-Vorpommern, which provided however high unemployment reduction benefits for the calculus.

In a study for the channel tunnel (by Vickerman 1989) the impacts from the point of investment and employment generation in various regions in England have been investigated.

The study of Vickerman (1989), quoted in the following, is based on an analysis of (i) the potential cost savings created by the Channel Tunnel (ii) the importance of transports in industry costs; and (iii) the potential for realising cost savings. The conclusion of the study was that there are "no compelling reasons for believing that the Channel Tunnel project will create an economic bonanza for the enjoining regions. If anything, there some evidence that benefits are more likely to accrue to locations at some distance from the Tunnel itself, say 100 to 150 km".

It is furthermore concluded that "there is little immediate gain to be made from the Channel Tunnel project either for the local economies at the tunnel-head or for the national economies".

This finally also concluded that the possible regional development effects, that could be generated, would only come about as a consequence of appropriate further complementary investments".

The Great Belt project has been subject to a considerable number of studies with various technics imployed. In a report from the Danish Transport Council the prediction of these studies are summarised as follows:

1. The effects on the port cities that will loose ferry operations will be negative;
2. The regional economic effects are asset as small;
3. The positive economic benefits that could be generated would accrue to a sizeable area from Copenhagen in the east to south-western Jütland.

For the Öresund link there are two studies available which are of a more qualitative nature. These studies pay emphasis on the lack of integration between the Swedish and the Danish economies in the vicinity of the fixed links at the present time, and therefore identify a considerable scope for further integration between the economies of the two countries flowing from the establishment of a fixed connection. In one of these studies the expected increase of the value of production in the agglomeration established through the fixed link is in the amount of 20 Bill. DKr. (Flyvbjerg, Bruzelius and Rothengatter, 1995).

But there is no detailed analysis given to present how this figure has been derived and whether this monetary value would be only due to the fixed link.

With respect to the Fehmarn Belt fixed link project there is a study available which comes out with a result that in the medium term there can not be substantial employment benefits due to increase economic activities or tourism. On the contrary, there is some concern of the counties adjacent to the fixed connection (Store Stroms Amt on the Danish side and Kreis Ostholstein on the German side) that the expected losses stemming from reduced ferry boat operations between Rodby and Puttgarden will not be offset by reallocation benefits from a fixed link. The long term corridor effects for the corridor Stockholm/Oslo/Malmö/Copenhagen/ Hamburg have not been analysed in detail.

What can be learnt from other large scale projects such as the Bosperus Bridge or the Faro River Bridge in Portugal is the following:

1. If serious capacity problems exist in the network in the region, investments to remove the bottlenecks are likely to affect investments and employment in the region;

2. Investments in large urban areas, where the new capacity for significant savings is in relation to cost of transport, are likely to give rise to relocation of both households (in the shorter run perspective) and of companies (in the longer term perspective); and

3. When there is a combination of various types of investments in infrastructure and social capital (investment in growth poles).

The most important lesson is that transport infrastructure investments will not automatically lead to an improvement of regional economic roads. The amount of regional effects is heavily independent on the economic environment, i.e. of existing bottlenecks and future complementary activities for regional development. This means that the forecast of regional effects can only be performed on the base of consistent and comprehensive scenario modelling including all regions and all networks affected in the system. There exist serious doubts whether a standardised procedure for evaluating regional growth effects as it is done by the German Ministry of Transport is helpful for decision making because the standard evaluation figures can create an absolutely wrong result on the real effects for the affected regions.

7 Project Appraisal and the Financing Scheme

A joint public and private partnership in project management and finance should in general be more goal driven and based on clear commitments of both partners. This implies that performance specifications for the project have to be set at the very beginning of the planning process including basic requirements for safety, environmental our spatial structure performance indicators. The role of project appraisal in such a scheme is not so much to give a once-and-for-all judgement on the viability of the project at the very end of the planning process but to provide a comprehensive evaluation of the projects performance according to the information available at the different stages of the planning process. In Figure 2 the structure for a project appraisal is outlined in a way that the cooperating public and private partners can extract the figures which they are interested in for preparing their strategic decisions. The figure shows that it is possible to classify the different effects of a project in a way that the financial rentability, the cost benefit ratio and the overall social balance of a project can be picked from one comprehensive appraisal sheet.

Valuation Method		Econo-mic					Regional/ Urban					Environm./ Safety					Valuation Criterion
		E1	E2	E3	E4	E5	R1	R2	R3	R4	R5	S1	S2	S3	S4	S5	
Monetary	dir	X	X	X	X	X											PRR
Evaluation $	ind							X			X	X	X	X			CBA
Quantitat. Evaluation *card. ut. index*									X						X		MCAC
Qualitative Evaluation *ord. ut. index*							X			X						X	MCAO

List of criteria

E	**Economic Impacts**
E1	Construction costs
E2	Maintenance costs
E3	Rolling stock and operation costs
E4	Revenue
E5	Generalised costs
R	**Regional and Urban Impacts**
R1	Concordance with regional land use planning (development axes)
R2	Contribution to regional economic balance
R3	Contribution to central gravity structure in space
R4	Improvement of urban settlement structure
R5	Improvement of the structure of employment sectors
S	**Environmental Sustainability and Safety**
S1	Safety
S2	Noise
S3	Local pollution of air, soil and water
S4	Global and strategic environment
S5	Land take and separation effects

Following from the financial prospects exhibited in the appraisal calculus the private investors can make up their decision whether they think to be able to finance a project completely by private funds or whether they need public support. The public sector can derive input data for calculating its maximal willingness for a publicly financed investment based on the evaluation of all externalities.

The forms of the public involvement in the financial scheme can be as follows:

- investment grants,
- interest guaranty,
- rentability guaranty,
- revenue guaranty.

Two ways are possible:

(1) Fix the public contribution before going for tender.

(2) Tender the project and check whether the best bid requires less public funds than the externality calculus has suggested to be the maximal public contribution.

In most cases a ceiling for a public investment grant, which is decided on before the private investors start their final evaluation, seems to be the best strategy. This is because by this way the risk which has to be taken by the private investors is most clearly defined and thus can be managed by the professionals while in the conventional approach the risk is almost totally allocated to the tax payer. Through this appraisal scheme the tax payer can feel more certain that he does not pay for the rent seekers but for externalities and distribution benefits, and partly for covering the regional development risk.

8 Conclusion

Private financing schemes with public partnership help to clarify the responsibilities of all partners involved. The interest of the public will be to control the externalities. Among these the regional impacts which can not be capitalised by private infrastructure investors play an important role. In many project appraisals these impacts are treated with very global and standardised instruments. This is not appropiate for a private finance approach for which a well founded figure on the external regional benefits has to be generated. Detailled regional impact analysis is necessary to provide the state authority with some baseline for the maximal public share of the total investment costs. Comparing the need for public funds which is exhibited in the private bid with the maximal public willingness to pay a final decision for the public participation can be taken.

References

Aschauer, D.A. (1989), Is Public Expenditure Productive? Journal of Monetary Economics. Vol. 23. 177-200.

Aschauer, D.A. (1990), Public Investment and Private Sector Growth. The Economic Benefits of Reducing America's "Third Deficit". Economic Policy Institute. Washington.

Bundesminister für Verkehr (1993), Gesamtwirtschaftliche Bewertung von Verkehrsinvestitionen. Schriftenreihe des BMV. Heft 72.

Danish Transport Council (1993), Fire baggrundsnotater til trafic 2005. Vol. 93/9. Copenhagen.

Flyvbjerg, B., N. Bruzelius and W. Rothengatter (1995), Fehmarn Belt: Issues for Accountability. Issued by the Danish Transport Council. Copenhagen.

Gramlich, E.M. (1994), Infrastructure Investment: A Review Essay. Journal of Economic Literature. Vol. 32. 1176-1196.

IFO (1990), Wettbewerbsfähigkeit ausgewählter EG-Regionen, IFO-Schnelldienst, Heft 9.

PROGNOS AG (1990), EG-Binnenmarkt und Beschäftigung. In: Materialien aus Arbeitsmarkt- und Berufsforschung. Nr. 5.

Vickerman, R.W. (1987), The Channel Tunnel and Regional Development. A Critique of an Infrastructure Growth Project. Project Appraisal Vol. 2. Not. Kent.

The World Bank (1994), World Development Report 1994: Infrastructure for Development. Oxford.

Institute for Regional Research
Publications since 1993

Beiträge aus dem Institut für Regionalforschung der Universität Kiel. Editor: Prof. Dr. Karin Peschel

13. *H. Herrmann, K. Ziesemer:*
    Zur Bedeutung und Entwicklung des Fremdenverkehrs im Kreis Ostholstein (1994).

14. *H. Herrmann:*
    Arbeitsmarktbericht für Schleswig-Holstein 1993, April 1994.

15. *A. Niebuhr:*
    Langfristiger Ausblick auf die Entwicklung der Wirtschaft in der Region Storstrøms Amt/Ostholstein, September 1994.

16. *M. Hirschfeld, R. Kersten:*
    Regionale Aspekte der Pflegeinfrastruktur in Schleswig-Holstein, Oktober 1994.

17. *R. Kersten:*
    Modelle eines regionalen Forschungstransfers am Beispiel der Christian-Albrechts-Universität zu Kiel - Befragung von Hochschullehrern, Mai 1995.

18. *H. Herrmann, M. Hirschfeld:*
    Arbeitsmarktbericht für Schleswig-Holstein 1994, Mai 1995.

19. *M. Hirschfeld:*
    Die wirtschaftliche Entwicklung Schleswig-Holsteins in der unmittelbaren Nachkriegszeit, Oktober 1995.

20. *H. Herrmann, A. Niebuhr:*
    Arbeitsmarktbericht für Schleswig-Holstein 1995. Februar 1996.

Diskussionsbeiträge aus dem Institut für Regionalforschung der Universität Kiel. Editor: Prof. Dr. Karin Peschel

28. *K. Schöler:*
    Zum strategischen Einsatz räumlicher Preistechniken. August 1993.

29. *D. Dohse:*
    Das Henry-George-Theorem. Dezember 1993.

30. K. Schöler:
Verhalten der Banken in oligopolitischen Märkten, 1994.

31. K. Schöler:
Einkaufsfahrten der Haushalte in einem räumlichen Marktmodell. 1994.

32. K. Schöler:
Ein räumliches Modell der Bank. 1994.

33. K. Schöler:
Tariffs, Factor Prices, and Welfare in a Spatial Oligopoly. 1995.

34. K. Schöler, H. Herrmann:
Leading Indicators for West German Regions. 1995.

35. K. Schöler, H. Sanner:
Räumliche Preisdiskriminierung in zweidimensionalen Wettbewerbs-märkten. 1995.

Scientific Reports and Expert Opinions since 1993

Commission of the European Communities, Dirctorate-General for Regional Policies (ed.): Trade and Foreign Investment in the Communities Regions: The Impact of Economic Reform in Central and Eastern Europe. Regional Development Studies, 7. Brussels, Luxembourg 1993. *Projektkoordination:* National Economic Research Associates, London. *Wissenschaftliche Leitung (Institut für Regionalforschung):* K. Peschel. *Bearbeitung (Institut für Regionalforschung):* J. Bröcker, O. Jäger-Roschko.

AKF Rapport,Sozioökonomische Analyse von Storstrøms Amt und des Kreises Ostholstein. Gutachten in Zusammenarbeit mit dem Amternes og Kommunernes Forskningsinstitut (AKF), Kopenhagen, im Rahmen des INTERREG-Programms der beiden Kreise. Kopenhagen 1994. *Wissenschaftliche Leitung:* N. Groes (AKF), K. Peschel. *Projektkoordination:* B. Madsen (AKF). *Bearbeitung am Institut für Regionalforschung:* M. Cnotka, H. Herrmann, A. Niebuhr, G. P. Radtke, K. Rupp, K. Ziesemer.

Pflegeberichterstattung 1994: Bestandserhebung über die Pflegeinfrastruktur in Schleswig-Holstein im stationären, teilstationären und ambulanten Bereich. Gutachten im Auftrag des Ministeriums für Arbeit, Soziales, Jugend und Gesundheit des Landes Schleswig-Holstein, Kiel 1994. *Wissenschaftliche Leitung:* K. Schöler. *Bearbeitung:* M. Hirschfeld, R. Kersten, H. Herrmann.

Arbeitsmarktbericht Schleswig-Holstein 1994, "Arbeitsmarktpolitik". Gutachten im Auftrag des Ministeriums für Arbeit und Soziales, Jugend und Gesundheit des Landes

Schleswig-Holstein, Drucksache des Schleswig-Holsteinischen Landtags, Kiel 1995. *Wissenschaftliche Leitung:* K. Schöler. *Bearbeitung:* H. Herrmann, M. Hirschfeld.

On potentials for long term development and policy options in the Baltic Sea region. Kurzstudie zur Vorbereitung des kooperativen Forschungsprojektes im Auftrag der Industrie- und Handelskammer zu Kiel. Kiel 1996. *Wissenschaftliche Leitung:* K. Peschel. *Bearbeitung:* S. Callsen, O. Jäger-Roschko, H. Herrmann.

Modelle eines regionalen Forschungstransfers am Beispiel der Christian-Albrechts-Universität zu Kiel. Forschungsprojekt gefördert von der Technologiestiftung Schleswig-Holstein: Teil I: Befragung von Hochschullehrern (Mai 1995). Teil II: Befragung von Unternehmensvertretern (April 1996). *Wissenschaftliche Leitung:* K. Peschel. *Bearbeitung:* R. Kersten

Essays and Papers since 1993

K. Peschel:

Perspectives of Regional Economic Development Around the Baltic. In: Visions and Strategies in European Integration, Hrsg.: L. Lundqvist, L.O. Persson. Berlin, Heidelberg, 1993.

H. Herrmann:

Die Beschäftigtenentwicklung in Schleswig-Holstein 1980 bis 1991 nach Berufsgruppen, Tätigkeitsbereichen sowie nach ihrer Qualifikation. In: W. Karr (Hrsg.): Küstenregionen im Strukturwandel, Beiträge zur Arbeitsmarkt- und Berufsforschung, Heft 169, Nürnberg 1993.

A. Niebuhr, H. Herrmann, K. Peschel:

Consequences of Economic Development in Eastern Europe for EU Regions. The Case of Storstrøm, Denmark, and Ostholstein, Germany. In: H. Cwiklinski, J.W. Owsinski (Hrsg.), Nordic-Baltic Europe is Restructuring: Looking for the Integration Fits., Warschau, Sopot 1994.

A. Niebuhr:

Die künftige Entwicklung in Osteuropa und ihr Einfluß auf Regionen des westlichen Ostseeraumes. Erschienen in U. Blien, H. Herrmann, M. Koller (Hrsg.): Regionalentwicklung und regionale Arbeitsmarktpolitik, Beiträge zur Arbeitsmarkt- und Berufsforschung, Heft 184, Nürnberg 1994.

H. Herrmann:

Wirtschafts- und Arbeitsmarktentwicklung in Schleswig-Holstein unter dem Eindruck der Grenzöffnung. In: U. Blien, H. Herrmann, W. Koller (Hrsg.): Regionalentwicklung und regionale Arbeitsmarktpolitik, Beiträge zur Arbeitsmarkt- und Berufsforschung, Heft 184, Nürnberg 1994.

K. Schöler:

Prognose kurzfristiger sektoraler Entwicklungen. In: Regionalentwicklung und regionale Arbeitsmarktpolitik, Beiträge zur Arbeitsmarkt- und Berufsforschung, Heft 184, Nürnberg 1994.

K. Schöler:

Verhalten der Banken in oligopolitischen Märkten. In: Jahrbuch für Sozialwissenschaft, Bd. 45, 1994, S. 389-403.

K. Schöler:

Zur Anwendung des Konzepts konsistenter konjekturaler Reaktionen auf die Theorie der räumlichen Preisbildung. In: H. Todt (Hrsg.), Beiträge zur Standortforschung, Schriften des Vereins für Socialpolitik, Bd. 227, Berlin 1994, S. 93-108.

J. Bröcker, O. Jäger-Roschko:

Eastern Reforms, Trade, and Spatial Change in the EC. Erscheint in: Papers in Regional Science, The Journal of the Regional Science Association International, 1995.

O. Jäger-Roschko:

Zukünftige Handelsverflechtungen im Ostseeraum. Projektzonen auf Basis eines Regressionsmodells. In: Raumforschung und Raumordnung im Ostseeraum, Arbeitsmaterial der Akademie für Raumforschung und Landesplanung, Nr. 215, Hannover 1995.

M. Hirschfeld:

Die wirtschaftliche Entwicklung Schleswig-Holsteins in der Nachkriegszeit. In: Aus der Geschichte lernen? Universität und Land vor und nach 1945. Eine Ringvorlesung im Wintersemester 1994/95, hrsg. vom Rektorat der Universität Kiel und dem Schleswig-Holsteinischen Landtag, Kiel 1995, S. 93-122.

K. Schöler:

Standorttheorien und Standortfaktoren (Art.). In: Akademie für Raumforschung und Landesplanung (Hrsg.), Handwörterbuch der Raumordnung, Hannover 1995, S. 923-926.

K. Schöler:

Räumliche Preistechniken im Wettbewerb. In: H. Todt (Hrsg.), Elemente der ökonomischen Raumstruktur. Schriften des Vereins für Socialpolitik, N.F., Bd. 237, 1995, S. 77-99.

H. Herrmann, K. Schöler:

Leading Indicators for West German Regions. Paper presented at the 22nd CIRET Conference, Singapur, September 1995. Erscheint 1996 im Konferenzband des CIRET (Centre for International Research on Economic Tendency Surveys), Hrsg.: K.H. Oppenländer und G. Poser.

You can get more information about the Institute for Regional Research's work, members and publications from the Institute's Activity Report. The report is available in English and German.

Institut für Regionalforschung der Christian-Albrechts-Universität zu Kiel

Institute for Regional Research at Christian-Albrechts-University in Kiel

Activity Report

Research and University Education at the Institute during 1993 and 1994

As of: August 1994

Mailing Address:
Institut für Regionalforschung der Universität Kiel, 24098 Kiel
Tel.: +49-431-880-3276, Fax: +49-431-880-3366
Email: ifr@bwl.uni-kiel.de, Internet: http://www.bwl.uni-kiel.de/vwlinstitute/ifr
Institute Office: Wilhelm-Seelig-Platz 1, 5th floor, 24118 Kiel

Contributions to Economics

Peter R. Haiss
Cultural Influences on Strategic Planning
1990. ISBN 3-7908-0481-9

Manfred Kremer/Marion Weber (Eds.)
Transforming Economic Systems: The Case of Poland
1992. ISBN 3-7908-0585-8

Marcel F. van Marion
Liberal Trade and Japan
1993. ISBN 3-7908-0699-4

Hans Schneeweiß/
Klaus F. Zimmermann (Eds.)
Studies in Applied Econometrics
1993. ISBN 3-7908-0716-8

Gerhard Gehrig/
Wladyslaw Welfe (Eds.)
Economies in Transition
1993. ISBN 3-7908-0721-4

Christoph M. Schneider
Research and Development Management: From the Soviet Union to Russia
1994. ISBN 3-7908-0757-5

Bernhard Böhm/
Lionello F. Punzo (Eds.)
Economic Performance
1994. ISBN 3-7908-0811-3

Michael Reiter
The Dynamics of Business Cycles
1995. ISBN 3-7908-0823-7

Michael Carlberg
Sustainability and Optimality of Public Debt
1995. ISBN 3-7908-0834-2

Lars Olof Persson/Ulf Wiberg
Microregional Fragmentation
1995. ISBN 3-7908-0855-5

Andreas Foerster
Institutional Analysis of Development Administration
1995. ISBN 3-7908-0853-9

Ernesto Felli/Furio C. Rosati/
Giovanni Tria (Eds.)
The Service Sector: Productivity and Growth
1995. ISBN 3-7908-0875-X

Giuseppe Munda
Multicriteria Evaluation in Fuzzy Environment
1995. ISBN 3-7908-0892-X

Boris Maurer
R & D, Innovation and Industrial Structure
1996. ISBN 3-7908-0900-4

Giovanni Galizzi/Luciano Venturini (Eds.)
Economics of Innovation: The Case of Food Industry
1996. ISBN 3-7908-0911-X

David T. Johnson
Poverty, Inequality and Social Welfare in Australia
1996. ISBN 3-7908-0942-X

Rongxing Guo
Border-Regional Economics
1996. ISBN 3-7908-0943-8

Oliver Fratzscher
The Political Economy of Trade Integration
1996. ISBN 3-7908-0945-4

Ulrich Landwehr
Industrial Mobility and Public Policy
1996. ISBN 3-7908-0949-7

Arnold Picot / Ekkehard Schlicht (Eds.)
Firms, Markets, and Contracts
1996. ISBN 3-7908-0947-0

Made in the USA
Las Vegas, NV
30 April 2026

46145986R00118